"Robert Zubrin has thought long and hard about travelling to Mars, so we should welcome his latest thoughts. In this highly readable book, he notes how technical advances, especially reduced launch costs, render such projects more feasible. He offers a fascinating and enticing vision of the utopian New World that he believes could and should be created on the Red Planet."

—MARTIN REESE,
the UK Astronomer Royal
and author of *On the Future*

"Zubrin challenges us to think beyond our current limitations and imagine a future where the Red Planet becomes a testament to our collective courage and resilience. A brilliant call to hope and to action!"

—JAMES R. HANSEN,
New York Times bestselling author of *First Man*

"At a time when an expanding presence in space raises important questions about how we operate there, it's refreshing to have this positive view on what we might achieve and what we might become as a civilization. Of course, our past has involved the bad, but that shouldn't obscure how we can extract the good to build a magnificent future society in the cosmos."

—CHARLES COCKELL,
author of *Interplanetary Liberty*
and *Taxi from Another Planet*

"An inspiring and informative vision of humanity's future on the Red Planet. Refreshingly optimistic and detailed, this exhilarating book takes readers on a vivid tour of our descendants' prospects on Mars."

—CHELSEA FOLLETT,
managing editor of HumanProgress.org
and author of *Centers of Progress*

THE
NEW WORLD
ON
MARS

WHAT WE CAN CREATE

ON THE RED PLANET

ROBERT ZUBRIN

DIVERSION
BOOKS

NEW YORK

Diversion Books
A division of Diversion Publishing Corp.
www.diversionbooks.com

Diversion Books and colophon are registered trademarks of Diversion Publishing Corp.

For more information, email info@diversionbooks.com

First Diversion Books edition, February 2024
Hardcover ISBN: 978-1-635-76880-0
eBook ISBN: 978-1-635-76995-1

Book design by Aubrey Khan, Neuwirth & Associates, Inc.

Printed in the United States of America
10 9 8 7 6 5 4 3 2 1

Diversion books are available at special discounts for bulk purchases in the US by corporations, institutions, and other organizations. For more information, please contact admin@diversionbooks.com

For Rachel

CONTENTS

THE
NEW WORLD
ON
MARS

1

WHAT CAN WE CREATE ON MARS?

We have it in our power to begin the world over again.
—THOMAS PAINE, *Common Sense*, 1776

HUMANS WILL SOON BE ABLE TO voyage to the Red Planet. This opens a truly grand question: What can we create on Mars?

Mars is the decisive step in humanity's outward migration into space. The Red Planet is hundreds of times farther away than the Moon, but it offers a much greater prize. Indeed, unique among the extraterrestrial bodies of our solar system, Mars is endowed with all the resources needed to support not only life but also the development of a technological civilization. In contrast to Earth's desert Moon, Mars possesses oceans of water in the form of huge glaciers and ice sheets, and it's frozen into the soil as permafrost. It also holds vast quantities of carbon, nitrogen, hydrogen, and oxygen, all in forms readily accessible to those clever enough to use them. Additionally, Mars has experienced the same sorts of volcanic and hydrologic processes that produced a multitude of mineral ores on Earth. Virtually every element of significant interest to industry is known to exist on the Red Planet. With its twenty-four-hour, day-night cycle and an atmosphere thick enough to shield its surface against solar flares, Mars is the only extraterrestrial planet that will readily allow large-scale greenhouses lit by natural sunlight.

For our generation and those to follow, Mars is the New World.

Mars appears barren to most people today, just as Ice Age Europe and Asia must have appeared to early humans migrating out of our original tropical African natural habitat. Yet, by developing new technologies, new attitudes, and new customs, our ancestors were able to create the resources to not only sustain themselves, but also to flourish with ever-increasing prosperity across the entire planet. In doing so, they transformed humanity from a local biological curiosity of Kenya's Great Rift Valley to a global family, hundreds of nations strong, spawning innumerable contributions to thought, literature, art, science, and technology.

Now the task before us is to multiply that triumph by developing the technologies and ideas that will enable the birth and sustain the growth of new, vibrant branches of human civilization on Mars, and subsequently on many worlds beyond.

The possibility of beginning the settlement of Mars within our time has recently been made apparent by the advances in space launch technology demonstrated by Elon Musk's SpaceX company. After a forty-year period from 1970 to 2010, during which the cost of a space launch remained static, the remarkable SpaceX team's introduction of mostly reusable launch vehicles has cut launch costs by a factor of five—from $10,000/kg to $2,000/kg—over the past decade.[1]

It is an extraordinary achievement. But now, even as I write these words, SpaceX is taking steps to slash costs by a further order of magnitude through the rapid development of a fully reusable two-stage to orbit heavy lift launch system they call Starship. The two stages together use about 5,000 tons of methane/oxygen propellant, but methane/oxygen only costs about $150/ton, putting total propellant costs per launch at $750,000. Starship will be able to deliver 100,000 kg of payload to orbit, and it expends no hardware. So, only $7.50

worth of propellant—and no hardware—will be consumed for every kilogram of payload sent to space. That means Starship could potentially cut Earth-to-orbit launch costs to under $100/kg—a cost that is two orders of magnitude lower than those prevailing a decade ago.

Critically, Starship's methane/oxygen propellant is the cheapest of all high-performance rocket propellants. Propellant costs are not important for expendable rockets because they are dwarfed by the cost of the hardware lost on each flight. But once rockets become reusable, fuel costs will matter as much as they do for airlines. The choice of a very inexpensive propellant combination opens the way for Starships to not only minimize launch costs, but also to develop vast new markets, such as surface-to-surface transportation, anywhere to anywhere on Earth in less than an hour. As a result, Starship's basic design is certain to be copied and produced in large numbers and diverse sizes by many companies from many nations. This competition will drive down costs even more, and eventually result in a market for used Starship-type vehicles, offering spaceflight capabilities for every budget.

There is another advantage to the Starship's propulsion system. Its propellant is readily manufacturable on Mars. The system is designed to enable human missions to the Red Planet.

A Starship mission to Mars would begin with the Starship lifted from the pad and fired out of the atmosphere with about one-third of orbital velocity by a booster stage called the Super Heavy. The Starship will then separate from the booster and continue to orbit using its own engines, while the Super Heavy returns to the launch site. Once reaching low Earth orbit, each Mars-bound Starship will need to be refueled by six more tanker Starships, putting the propellant cost of sending a Starship to Mars with a hundred tons of payload, including a hundred passengers, at around $5 million. That works out to $50,000 per passenger—or less if part of the cost is paid by those shipping the freight. If SpaceX can increase the Starship production rate to one per week—they are already producing prototype units at a rate of one per month—they can probably cut the cost of building a Starship to something in the $10 million range. (The very large booster first stage, called the Super Heavy, would cost more, but, as it always returns to the launch site the same day it takes flight, only a few Super Heavies would be needed to support a large Starship fleet.) This means, if passengers were charged $300,000 each—or $30 million total for a hundred people—that would be enough to pay for the Starship (the immigrants would be wise to keep it for housing), its cargo, and the launch operations and propellant to get it to Mars, with a handsome profit for SpaceX.

Now, $300,000 for a ticket to Mars, plus tools, provisions, and a starter apartment in a landed Starship, is an interesting figure. It is the equivalent, in modern terms, of what it cost to travel from England to colonial America in the 1600s. At that time, a middle-class person could pay for his family's one-way passage by selling his house and farm, while a workingman could get his ticket in exchange for seven years' of labor for room and board. Roughly speaking, that's equivalent to what a $300,000/ticket represents today. It's a price that a sizable number of people could manage if they were willing to cash in their chips, pull up stakes, and take a chance on a new life in a new world.

In short, people will soon be able to get there. So, again, *what will we create on Mars*?

NASA is interested in exploring Mars for scientific purposes. That is a legitimate motive for the space agency. The early Mars was very much like the young Earth, a warm, wet planet with a carbon dioxide–dominated atmosphere. We know from fossils and other biomarkers that microbial life appeared on Earth 3.7 billion years ago—virtually as soon as the planet had cooled enough to allow liquid water. But what we don't know is whether the processes that led to the rapid appearance of life on Earth were a sure bet based on a sequence of natural complexification of molecules endemic to chemistry, or whether it was a matter of freak chance. If it was the former, then life is everywhere in the universe—including almost certainly intelligent life, since once life gets started, evolution will take it in all possible directions. But if it was the latter, then we could be virtually alone.

We can find the answer to this question by going to Mars. If life ever prevailed in the now long-gone seas of Mars, it will have left fossils behind, which could be found by human rock hounds. Not only that, while the planet's cold desert surface is hostile to aquatic microbes, survivors of such ancient life could still exist in reservoirs of geothermally warmed liquid water that we now believe to exist underground on Mars. Human explorers on Mars could set up drilling rigs and bring up samples of such water. If the samples contain life, we will not only be able to detect it but learn about its biochemistry.

Life on Earth exhibits great diversity, but it all uses the same system for recording and transmitting information from one generation to the next. Whether bacteria, mushrooms, pine trees, grasshoppers, crocodiles, or humans, they all utilize the identical DNA/RNA information system. In this respect, biologists today are like linguists who are only acquainted with English, completely unaware of other languages, let alone other alphabets like Chinese, which employ totally different systems for transmitting thought.

But if we find Martian life, we can begin to deepen our understanding of life itself. We can find out whether life as we know it on Earth is what life *is*, or alternatively, is just one example drawn from a vastly greater tapestry of possibilities.

The implications of discovering an alternative biological information system would be beyond reckoning, and not just for pure science, but for new types of biotechnology that could revolutionize human existence in the twenty-first century.

Mars is the Rosetta Stone for revealing the truth about the potential prevalence and diversity of life in the universe, something that thinking women and men have wondered about for thousands of years. The search for that truth is well worth risking life and treasure.

But, as important as this search is, Mars is not just an object of scientific inquiry. It is a world with a surface area equal to all the continents of Earth put together, containing all the materials needed to support not only life but technological civilization.

Can We Create New Civilizations on Mars?

The question of colonizing Mars is not fundamentally one of transportation. If we were to use a vehicle comparable to the SpaceX Starship now under development to send settlers to Mars on one-way trips, firing them off at the same rate SpaceX is currently launching its Falcon rockets, we could populate Mars at a rate comparable to the rate at which the British colonized North America in the 1600s—and at much less expense relative to our resources. From a technical point of view, the problem of colonizing Mars is not that of moving large numbers to the Red Planet, but of the ability to create and use Martian resources to support an expanding population once we are there. The technologies required to do this will be developed at the first Mars base, which may well be established for scientific purposes, but which will also act as the beachhead for waves of immigrants to follow. To reduce its logistics requirements, the first base will inevitably seek to develop techniques for extracting water from the soil, for conducting increasingly large-scale greenhouse agriculture, for making ceramics, metals, glasses, and plastics out of local materials, and constructing ever larger pressurized structures for human habitation and industrial and agricultural activity.

Over time, bases could transform themselves into small towns. The high cost of transportation between Earth and Mars will create a strong financial incentive to find astronauts willing to extend their surface stay beyond the basic eighteen-month tour of duty, to four years, six years, and longer. Experiments have already been done showing that plants can be grown in greenhouses filled with carbon dioxide at Martian pressures—the Martian settlers can thus set up large inflatable greenhouses to provide the food required to feed an expanding resident population. Mobile units can be used to extract water from Mars's abundant ice and permafrost, supporting such agriculture and making possible the manufacture of large amounts of brick and concrete, the key materials required for building large, pressurized structures. While the base could initially comprise an interconnected network of ships, "tuna can" habitats, or other flight systems, by its second decade, the base's settlers might live in pressurized, underground-vault domains the size of shopping malls. Not too long afterward, the expanding local industrial activity will make possible a vast expansion in living space by manufacturing large supplies of high-strength plastics like Kevlar and Spectra that could allow the creation of inflatable domes encompassing sunlit, pressurized areas up to hundreds of meters in diameter. Each new nuclear reactor landed will add to the power supply, potentially augmented by locally produced photovoltaic panels and solar thermal power systems. However, because Mars has been volcanically active in the geologically recent past, it is also highly probable that underground hydrothermal reservoirs exist on the Red Planet. Once such reservoirs are found, they can be used to provide the settlers with abundant supplies of both water and geothermal power. As more people arrive in steady waves and stay longer before they leave, the population of the towns will increase. In the course of things, children will be born and families raised on Mars, the first true colonists of potentially many new branches of human civilization.

We don't need any fundamentally new or even cheaper forms of interplanetary transportation to send the first teams of human explorers to Mars. However, meeting the logistical demands of a Mars base will create a market that will bring increasingly lower-cost commercially developed systems for interplanetary transport. Combined with the base's own activities in developing Martian resources, such transportation systems will make it possible for the large-scale colonization of Mars to begin.

Technical feasibility alone is insufficient to enable settlement of the planet. While the initial exploration and base-building activities can be supported by government, nonprofit society, or corporate largesse, a true colony must eventually become economically self-supporting. Mars has a tremendous advantage compared to the Moon and asteroids in this respect, because unlike these other destinations, the Red Planet contains all the necessary elements to support both life and technological civilization, making self-sufficiency possible in food and all basic, bulk, and simple manufactured goods.

That said, Mars is unlikely to become autarchic for a very long time, and even if it could, it would not be advantageous for it to do so. Just as nations on Earth need to trade with one another to prosper, so, too, the planetary civilizations of the future will need to engage in trade. In short, regardless of how self-reliant they may become, the Martians will always need cash. Where will they get it?

It may be possible to export plentiful Martian deuterium (fusion power fuel) to Earth. An even more lucrative business will be exporting food and other necessaries to miners in the asteroid belt. Many main belt asteroids contain large amounts of platinum group metal ores far richer than any that exist on Earth. The development of inexpensive spaceflight will consequently open the way to an asteroid gold rush. But the surest way to make money off a gold rush is not to mine gold, but to sell blue jeans to gold miners. Because of its lower gravity and positional advantage, it will be about a hundred times cheaper to send supplies and equipment to the asteroid belt from Mars than from Earth. Anything the miners need that can be made on Mars will be made on Mars. What San Francisco was to the forty-niners, Mars will be to the belt.

In my view, however, the most important of all Martian exports will be patents. The Mars colonists will be a group of technologically adept people in a frontier environment where they are free to innovate—indeed, *forced* to innovate—to meet their needs, making the Mars colony a pressure cooker for invention. For example, the Martians will need to grow all their food in greenhouses, strongly accentuating the need to maximize the output of every square meter of crop-growing area. They thus will have a powerful incentive to engage in genetic engineering to produce ultra-productive crops and will have little patience for those who would restrict such inventive activity with fearmongering or red tape.

Even more significant, there will be nothing in shorter supply in a Mars colony than human labor time. Just as the labor shortage in nineteenth-century America drove Yankee ingenuity to create a series of labor-saving inventions, the labor shortage on Mars will drive Martian ingenuity in such areas as automation, robotics, and artificial intelligence. Recycling technology that recaptures valuable materials otherwise lost as waste will also be advanced. Such inventions, created to meet the needs of the Martians, will prove invaluable on Earth, and the relevant patents, licensed on Earth, could produce an unending stream of income for the Red Planet. Indeed, if the settlement of Mars is to be contemplated as a private venture, the creation of such an inventors colony—a Martian Menlo Park—could conceivably provide the basis for a fundable business plan.

Martian civilizations will become rich because their people will have to be smart. They will benefit Earth not only as fountains of invention, but as examples of what human beings can do when they rise above their baser instincts and invoke their full creative powers. They will show all that infinite possibilities exist, not to be taken from others, but to be made.

This is an important point. There are really no such things as "natural resources." There are only natural raw materials. It is human creativity that transforms raw materials into resources through technological innovations. On Earth, land was not a resource until people invented agriculture. Oil was not a resource until people invented petroleum drilling and refining and machines that could run on the products. Aluminum was not a resource until the late nineteenth century, when technologies were invented to extract the metal from aluminum oxide. Until then, it was just dirt. Uranium was not a resource until nuclear power was developed. Deuterium is not a resource now, but it will be once fusion power is invented. It is not the Earth, but people, who have created all the resources that sustain us.

Mars has no resources now, only materials. But it will have abundant resources once resourceful people are there. All Mars really needs is people. But why would they move there?

Help Wanted

One reason people will immigrate is because of the labor shortage itself. Labor shortages may be tough for businessmen, but they are great for workers. Labor shortages mean high pay. While conditions in colonial, frontier, and nineteenth-century America may seem harsh to us today, they offered—by comparison—fantastically high compensation and unprecedented opportunities to poor people from Old World Europe and Asia. So, they voted with their feet, by the millions, to bet all their savings and risk perilous voyages, leaving behind all they had known to cross the ocean to a new world.

Men and women do not live by bread alone. If a Mars city is to succeed and grow, it will need to be a place to which all kinds of people want to move. Few will want to immigrate to Mars to live in dingy environs. A Mars base may only need to be functional, but a Mars city needs to be beautiful. If it is to succeed, it will need to pay ample attention to aesthetics. But it is not only physical beauty that makes a city joyful; it's the opportunity a city offers citizens to develop and exercise their full human potential. What social, political, and cultural forms might a city create to make it a land of wonders, dreams, freedom, and opportunity to draw millions of hopeful immigrants of every class and talent?

Martians will have to come up with clever technical, economic, and aesthetic solutions to the problems of designing a practical and beautiful Mars city-state. What will ultimately be decisive, however, is the ability of Martians to define a better way for humans to live together. Would-be colonists will no doubt differ on what the foundations of such a society might be, with candidates that seem reasonable to us today running the gamut from social democratic to libertarian. The Martians will no doubt conceive many more. Hence, there will undoubtedly be not one, but many Martian city-states created from alternative visions of what makes for a better life.

I see such diversity of possibilities as a great strength. Mars is big enough for numerous colonies, founded by a wide variety of people who will have their own ideas as to what forms of social organization offer the greatest scope to realize human hopes and full potential. Indeed, the chance to be a maker of one's own world, rather than just an inhabitant of one already made, is a fundamental form of freedom, whose attraction may well prove to be the primary driver for many to accept the risks and hardships that settling another planet must necessarily involve. It is hardly to be imagined that human social thought has reached the final and best answers possible in the early twenty-first century. There will always be people with new ideas who need a place to go where the rules haven't been written yet—a place to give their ideas a fair try. That said, not all their notions will be workable. Some may prove impractical and cause the colonies that adopt them to stagnate or fail altogether. But sometimes a new path can lead upward.

Whether they wish to or not, Martian cities will compete for immigrants. The ones with the best ideas will draw the most people. This is why dystopian totalitarian space colonies controlled by villains who tyrannize their subjects by threatening to cut off their air will remain mere fictions. A successful extraterrestrial tyranny is impossible because no one would move there.

Again, the fundamental reality that will shape the formation of Martian civilizations will be the labor shortage. Martian cities will neither be willing nor able to block the exercise of talent by creating artificial certification obstacles to job or profession entry, or throw people away, as we currently do when we stash our elderly in old-age homes. Nor will they be able or willing to enforce prolonged childhood on adolescents by institutionalizing them until their third decade in schools. Rather than being shunned, immigrants and new arrivals of every kind will be valued and welcomed. These latter will most emphatically include children.

Mars needs children. That means Mars needs women.

In the seventeenth and eighteenth centuries, Britain and France vied for control of North America. France was the much larger power, with a population four times that of Britain. Yet while British colonists built real communities in North America, the French built fur trading posts. Consequently, British women were willing to move to America, but French women were not. As a result, by the time the struggle between the two powers reached its climax in the 1750s, there were two million English-speaking colonists in North America but only fifty thousand French. The outcome was inevitable, with results that shape the world to this day.

Mars will have its share of rough outposts. But only places that draw women will develop into civilizations, and the cities that attract the most female immigrants will outgrow the rest. How they will go about doing this will be theirs to devise or hit upon by chance. Suffice to say, the status of women in the ultimately prevalent Martian societies is likely to be high.

Biological evolution occurs through natural selection. Human social evolution does so as well, except that innovations need not be arrived at randomly, but can be *chosen*, and if found successful, transferred, and widely propagated by thought. This makes the kind of social experimentation that will occur on Mars a far more powerful force for progress.

The ideas of eighteenth-century enlightenment liberalism that America's founders used to initiate their new republic were not original with them. On the contrary, they were well known to educated circles in Europe at the time. But those who believed in such concepts as government of the people, by the people, for the people, freedom of speech, freedom of worship, and inalienable individual rights that stand above king, state, or church were dismissed as impractical dreamers, perhaps pleasantly insane. It took a new beginning in a new world—a place where the rules hadn't been written yet—to give these ideas a chance to show what fruit they might yield. America's founders called their project a "noble experiment," as indeed it was. To be sure, it did not produce perfect results, but it was the best the world had to offer at the time. And this experiment did well enough that millions of people voted with their feet to come to America to be part of it and help build America into a beacon of hope, a powerhouse of creativity, and a shining example to be developed further by and for all humankind.

There are likely to be many noble experiments on Mars. Some will fail, but others will succeed, leading their cities to grow, prosper, invent, create, and, by example, set a new standard for the further progress of humanity everywhere.

We need to keep looking for something better. Those who search may find. What could be more important? In taking on the challenge of Mars, we hold it in our power to begin the world anew.

What a grand thought. What a grand opportunity. Let us prove worthy of our moment.

2

A BRIEF HISTORY
OF MARS

IF YOU WANT TO EXPLORE MARS, you really have to go there.

The Renaissance German astronomer Johannes Kepler used observations of Mars to divine the laws of orbital motion, thereby proving that Earth was a planet. By implication, therefore, the planets, those little moving lights in the sky, were really vast worlds like the Earth. But how to explore these incredible, newly recognized bodies?

Kepler thought we should go. As he wrote to his Italian counterpart Galileo Galilei in 1609, "Ships and sails proper for the heavenly air should be fashioned. Then there will be people who do not shrink from the dreary vastness of space."

It was a great idea, but a bit premature. However, Galileo had taken a curious Dutch optical invention and improved it to create an effective instrument of naval warfare. For his telescope, he was handsomely rewarded by the sea lords of Venice. Now he turned it toward the heavens and found wonders galore. He discovered mountains on the Moon—making it plain to see that it was indeed a world. He also found four moons orbiting Jupiter, giving additional credence to the Keplerian view of the universe, and providing European navigators with a celestial clock, enabling the calculation of longitude and the drawing of the first accurate maps of the Earth. Soon enough, other telescopes were trained on Mars.

The Italian astronomer Francesco Fontana produced in 1636 the first drawing of Mars through the telescope, though viewed today it reveals no recognizable features. In 1659, the Dutch astronomer Christiaan Huygens produced the first drawing that shows a known Martian feature, a roughly triangular dark blotch that appears on the planet's face, today known as Syrtis Major. By carefully observing Syrtis and similar features, early astronomers determined that the Martian day, or sol, was similar to Earth's. In 1666, the Italian Giovanni Cassini measured the Martian day at twenty-four hours, forty minutes, and about two and one-half minutes—longer than today's accepted measure of twenty-four hours, thirty-seven minutes, twenty-two seconds. Although Cassini was also apparently the first to note one of Mars's polar caps, Huygens in 1672 produced the first sketch of one of the caps. Utilizing observations made between 1777 and 1783, William Herschel, discoverer of Uranus, noted that Mars should have seasons, as its polar axis was tilted about 30° (25° is the modern value) to its orbital plane.

Observations of Mars continued through the decades, especially around "oppositions," those times when Mars (technically, any planet outside Earth's orbit) lies on the opposite side of the Earth from the Sun. At these times, Mars is at its closest to Earth and thus shines its brightest. By the early nineteenth century, astronomers had collected a basketful of basic Mars statistics: its orbital period, the length of its day, the planet's mass and density, distance from the Sun, and surface gravity. But what truly intrigued observers was the changing face of Mars. Through the years, the telescope's eyepiece revealed that Mars's face was mottled with dark patches that came and went with time. Likewise, the bright white spots observers noted at the poles appeared to vary with the Martian seasons, expanding and contracting over the course of a Martian year. And Mars apparently hosted an atmosphere, as some observers spied vague indications of clouds above the Martian surface.

The opposition of 1877 proved especially fruitful for observers and for Martian studies. In his work *Gulliver's Travels*, published in 1726, the great British satirist Jonathan Swift speculated that Mars must have two moons, because Earth has one and Jupiter has four. In 1877, Asaph Hall of the United States Naval Observatory found both, and promptly named them Phobos and Deimos—fear and terror, an appropriate entourage for the planet of war.

But the big Mars news of the 1877 opposition came from the Italian astronomer Giovanni Schiaparelli, director of the Brera Observatory in Milan. Schiaparelli's reports of his observations noted the location of more than sixty features on the Martian surface. Along with many standard features, he reported sighting linear markings crossing the face of Mars. He named these features after terrestrial rivers—Indus, Ganges—but referred to them in his writings as *canali*. While not the first to note these strange markings, he was the first to identify an extensive system of *canali*.

The Italian word *canali* is ambiguous. It can translate as simply "channels" or "grooves." But it can also mean "canals." The wealthy Boston Brahmin Percival Lowell decided upon the latter interpretation.

Born into an illustrious New England family of poets, educators, statesmen, and industrialists (the great poet Amy Lowell was his sister; his brother, Abbott, was president of Harvard), Lowell while in his late thirties became intrigued with Mars, especially with Schiaparelli's observations. No wonder. Canals reflect the work of minds in collaboration, of life. Lowell decided that Mars demanded his attention, and devote his attention to Mars he did, with a passion and pocketbook few could match.

The tool Lowell built for his investigations—the Lowell Observatory in Flagstaff, Arizona—saw first light in April 1894, just a few weeks before Mars reached its biennial opposition with Earth. Lowell and his staff atop Mars Hill spent more than a decade studying and mapping the face of Mars, including hundreds of canals. In their number and organization, Percival Lowell saw the history of an alien race trying to survive on an arid, dying world plainly writ.

Lowell captured the popular imagination with his sympathetic picture of an intelligent race of Martians trying to forestall its inevitable doom. The effect of his writings was amplified further by adventure writers such as Edgar Rice Burroughs, who used the Lowellian Mars as the setting for an extraordinary and romantic Martian civilization that called its home planet "Barsoom." Burroughs's Mars novels featured swashbuckling heroes rescuing daring and beautiful princesses endangered by monsters, savages, and power-mad Martian tyrants, all set against a rich tapestry of life on Barsoom. In its Barsoomian incarnation, Lowell's Mars enchanted millions of readers.

Over the years, though, the tide of opinion slowly turned against Lowell as other observers using more powerful telescopes found no evidence whatsoever of canals. We now know that Lowell was wrong in his investigations of Mars, but he did leave an important legacy behind: He fired the imaginations of people to make them see a world on Mars. It was Lowell's works that inspired the pioneers of rocketry, including Robert Goddard and Hermann Oberth, to begin their quest to develop the tools that would soon make the solar system accessible, not only to the eye, but to the hand of man.

By the 1950s, the canals were gone from the scientific literature, and all that remained was Mars's orbital and rotational data, its gravity, an estimate of its surface temperatures, and the facts that it had an atmosphere, dust storms, polar caps that might be water or dry ice, and two moons. There were also observations of seasonal surface color changes, which some observers still hoped might represent the annual advance and retreat of simple vegetation, while others ascribed them to dust or dismissed them entirely.

There matters stood until October 1957, when little Sputnik took off into orbit and announced to an amazed world with its spunky radio beep-beep that a new era in human history had begun.

With the advent of the space age, we would no longer be limited to gazing on Mars from afar. Soon enough, we would be able to go there and see for ourselves. But with enlightenment came disappointment.

Though Lowell's visions had long since been discarded, the idea that Mars might harbor some form of life had never died. Streaking by the planet in July 1965, the first spacecraft to visit Mars, American *Mariner 4* quashed once and for all the Lowellian vision of the Red Planet, revealing a barren, cratered surface, more Moon-like than Barsoom-like. During the summer of 1969, *Mariner 6* and *7* confirmed their predecessors' findings. Science experiments confirmed *Mariner 4*'s atmospheric findings—the pressure of the carbon dioxide-rich atmosphere was low, just six to eight millibars. (A millibar is 1/1,000th of Earth's sea-level atmospheric pressure, so at seven millibars, Mars's atmosphere was a bit less than 1 percent as thick as Earth's.) Temperatures measured near the south pole supported the notion that frozen carbon dioxide—dry ice—formed the polar cap. Mars, according to the *Mariner* flybys, was a cold, dead, cratered planet—not a place you want to linger. Then came *Mariner 9*.

Unlike the previous American spacecraft, *Mariner 9* would go into orbit around Mars. Where the early *Mariners* shot by the Red Planet and captured what information they could, *Mariner 9* and a companion spacecraft would map the planet's surface and observe planetary dynamics over a sixty-day period. Unfortunately, that companion spacecraft, *Mariner 8*, ended up in the waters of the Atlantic shortly after launching in the spring of 1971. *Mariner 9*, though, lifted off flawlessly on May 30, bound for Mars. Just days earlier, the Soviet Union had launched *Mars 2* and *Mars 3*, combination orbiter/lander spacecraft.

On September 22, about two months before the *Mars* probes and *Mariner* were due to arrive on the Red Planet, astronomers noticed a bright white cloud begin to develop over the Noachis Terra region of Mars. The cloud grew quickly, by the hour. Within days, the cloud, by then recognized as a dust storm, had enveloped the planet. As robotic eyes sped toward Mars, the planet pulled a shroud around itself. Far-encounter photographs of the planet captured by *Mariner 9* on November 12 and 13 showed a blank disk, save for a slight brightening near the south pole and a few small, dark smudges above the equator. On November 14, five and a half months after leaving Earth, the spacecraft slipped into Mars orbit to gaze down on an essentially featureless planet. The probe's controllers rewrote the mission plans, allowing for some science experiments and photography to be undertaken, but, in essence, told the spacecraft to kick back and ride out the storm.

Mars 2 and *3* didn't have that option. Unlike *Mariner*, the Soviet program did not have adaptive operational capability. Each probe was outfitted with a lander and an orbiter. On arrival at Mars, the orbiters duly released their landers into the maw of the largest Martian dust storm ever recorded. Parachuting blindly through an atmosphere whipped by winds of 160 kilometers per hour (about 100 miles per hour), both landers hit the ground too hard for their airbag deceleration systems to save them. The *Mars 2* lander was destroyed on impact; the *Mars 3* lander managed to transmit twenty seconds of data after crashing, and then died.

The Soviet orbiters hardly fared any better than the descent probes. Nearly all data from the *Mars 2* orbiter was lost because of poor telemetry, and the *Mars 3* orbiter pulled into a wildly elliptical orbit around Mars, producing only one released photograph.

While the dust storm raged, and the Soviet probes met their respective fates, *Mariner 9* serenely orbited the planet, waiting for the dust to clear, both literally and figuratively. Toward the end of December and into early January 1972, the Martian skies started clearing, and *Mariner 9* began to return staggeringly vivid images of an unimagined world.

The small smudges *Mariner* imaged during its far encounter could now be seen for what they were: enormous mountains whose tops *Mariner* had spied through the dust storm. A century earlier, optical astronomers had noted a bright region in the largest of these massifs, and dubbed the region Nix Olympica, "the Snows of Olympus." It was an apt name, as Nix Olympica proved to be the largest mountain in the solar system—Olympus Mons—looming some twenty-four kilometers (fifteen miles) above the Martian surface and covering an area about the size of the state of Missouri. Another region of Mars well known to astronomers, the Coprates quadrangle, yielded surprises as well. Through the telescope, Coprates appeared as a stubby, bright, cloudlike band. As skies cleared, *Mariner*'s audience of scientists realized they were looking at a dust cloud slowly settling into the bottom of a valley of mind-boggling immensity. Now known as Valles Marineris (in honor of *Mariner 9*), this ragged scar stretches nearly four thousand kilometers (about two housand four hundred eighty five miles) across the planet. Up to two hundred kilometers wide and six kilometers deep, Valles dwarfs any similar feature on Earth (you could tuck the Rocky Mountains in one of Valles's side valleys and nobody would see them).

With each orbit of the planet, *Mariner* returned ever more astonishing information. The greatest surprise, though, proved to be images of sinuous channels (yes, *canali!*) that appeared to have been carved by running water—there were riverbeds on Mars.

Whatever romance the earlier *Mariners* had killed, *Mariner 9* renewed. The probe reinforced many of the earlier *Mariner* findings, but overturned others, including the notion that Mars was simply a knockoff of the Moon. Imagine the Martian globe bisected by a line running at roughly a 50° angle to the planet's equator. Below that line to the south lies the heavily cratered, ancient terrain *Mariners 4, 6,* and *7* discovered and recorded. North of the line, craters are few while evidence of more recent geological activity is plentiful. It just happened that the first three *Mariners* visited the South, offering no clues as to what other regions of the planet might reveal. *Mariner 9*'s images (more than seven thousand of them) and data swept away the notion of the Red Planet as a "cosmic fossil." Instead, *Mariner 9*'s findings told the tale of a planet of fire and ice. In the distant past, Mars's surface had been geologically alive. Volcanoes had roared and resurfaced vast areas of terrain; internal mechanisms of some sort had fractured and split the landscape, lifting the Tharsis region (on which Olympus Mons stood) kilometers above the landscape; and water had flowed across the planet's surface in volumes large enough and for periods long enough to carve the face of the planet. Mars was once warm, wet, and alive with geologic activity. And that begged the question once again: Was Mars now, or perhaps in the past, bustling with biologic activity, with life?

To answer that question, NASA sent the *Viking* mission to Mars.

The *Viking* program was straightforward in description—two orbiters, two landers, all to head for Mars in 1973 to search for life— but proved staggeringly difficult in execution. A budget squeeze delayed launch until 1975, which, in retrospect, was a hidden blessing, as the spacecraft simply would not have been ready by 1973 without, in the words of a *Viking* team member, "compromising both capability and reliability."

The four *Viking* spacecraft bristled with instruments for imaging, water-vapor mapping, thermal mapping, seismology, meteorology, and more, but the heart of the mission lay with the landers' biology packages. *Viking* engineers had packaged three biology labs weighing about nine kilograms total into something that could sit quite comfortably in your bookcase.

The three experiments in the biology package operated on the same basic principle: seal some Martian dirt in a container with a culture medium, incubate it under different conditions, and then measure the gases emitted or absorbed. The experiments differed in the specific approaches they took to incubate samples and in what they sought to detect and measure as evidence of life. The *Viking* landers also carried an X-ray florescence instrument capable of assessing the elemental composition of the soil, and a gas chromatograph mass spectrometer (GCMS) capable of detecting and identifying organic compounds in the soil.

The search for life began on *Viking 1*'s eighth Martian day—sol 8 in the local time zone; July 28, 1976, here on Earth—as the lander extended its sampler arm, dragged it across the Martian surface, and delivered soil to the biology package. The three experiments received their small allotments of soil and set to work. Over the course of the next three days, incredibly, all three biology experiments reported powerful gas releases, positive signals for life, in some cases virtually immediately after exposure of the culture media to Martian soil.

The *Viking* biology team was, to say the least, stunned. Three experiments, three positive responses, three indications of life . . . maybe. The gas release signals were definite, but the suddenness of both onset and cessation had the ring of chemical reaction more than biological growth. So, caution was called for. The discovery of life anywhere in the solar system would have profound ramifications not just for the world of science but for the entire world community. Once again, as in Kepler's time, humanity would come to know its place in the universe more fully, more truthfully. We would know that while we are not the center of the universe, we are part of a phenomenon that is general throughout the universe. We would know that life owns the universe. This was, most definitely, no small announcement.

No one on the biology team was eager to rush out such an announcement, only to discover that he had jumped the gun. So, conservatism prevailed, especially since many on the biology team had strong suspicions that the reactions witnessed were nonbiological in origin.

On sol 23, the gas chromatograph mass spectrometer (GCMS) analyzed a sample of Martian soil and found not a trace of organic carbon. After the reactions recorded by the three biology packages, this came as an enormous surprise and heightened the debate. Scientists had expected the GCMS to find at least some trace of organic compounds of nonbiological origin, such as materials from meteorites. In fact, that was a concern surrounding the GCMS—how to tell biologic organics from nonbiologic. But now, with the GCMS recording absolutely no evidence of organics in Martian surface soils, the search for life on Mars became for some a search for processes that could reconcile the discovery of an evidently lifeless Mars with the biological results.

On September 3, *Viking 2* settled down on the Utopia Planitia, nearly halfway around the planet, some 6,400 kilometers from the *Viking 1* landing site and about 25° farther north. The biology experiments and the GCMS were soon up and running, investigating soils that appeared to be slightly moister than samples from the Chryse site (30 degrees north, 300 degrees east). Again, results from the biology experiments gave positive responses that appeared to be more indicative of chemistry to some, and the GCMS found no trace of organic carbon. Again, the results caused a stir, with some investigators holding out for biology, others chemistry. Again, the results highlighted a basic problem: The *Vikings* could perform four experiments and only four, and three were saying "looks like life," while the other was saying "no way." If the soil samples had been in a terrestrial lab, dozens of additional experiments could have been performed to resolve the argument definitively. On Earth, the samples could even have been incubated in a culture medium and the results observed directly with a microscope. But in *Viking*'s limited, four-experiment lab on Mars, none of this was possible. In essence, we were left with contradictory results. In the words of writer Leonard David, "*Viking* went to Mars and asked if it had life, and Mars answered by replying 'Could you please rephrase the question?'"

The results thus remain controversial to this day. My own feeling is that the *Vikings* did not detect life in the surface soil of Mars, because there is no liquid water there and virtually no organics. So, while abstract arguments can be made for a "sparse spore" hypothesis, it seems almost impossible to construct a rational theory explaining how the life cycle of these putative Martian surface organisms would function. Furthermore, since Mars has very little in the way of an ozone layer in its atmosphere, the surface is bathed in ultraviolet light of sufficient intensity to do a pretty good job of sterilizing the planet's surface of microorganisms. However, this does not rule out the possibility of microbial life below the surface. Indeed, there are families of bacteria known as chemotrophs that derive their energy from various inorganic chemicals as opposed to sunlight (as plants do) or organic nutrients (as we do). A small group adapted to temperatures of 70° to 90° centigrade, and living happily by oxidizing sulfur for their energy requirements, would probably feel right at home in some underworld environments that, as more recent discoveries now show, almost certainly exist on Mars.

All creatures great and small surviving in extreme environments have one thing in common: Their environment includes a source of water, however meager. The fact that Mars shows a remarkable amount of evidence of both surface and subsurface water in its distant past argues for the possibility of life in the past or perhaps even now. Host environments to such life could be thermal hot spots, such as hot springs, subsurface hot spots, a subsurface water table, subsurface or near-surface brines, subsurface permafrost, or perhaps even areas with evaporite deposits, such as the salt formations that have been found to be home for millions of years to earthly bugs. In the 1990s, investigators in the state of Washington discovered a species of bacteria living deep underground, subsisting on the chemical energy derived from the reaction of cold groundwater with basalt. There does not seem to be any reason for believing that similar organisms could not survive equally well in the subsurface environment that is believed to exist on Mars. The point is that life is tough and finds ways to live to the very limit of the possible. No one expects to find herds of six-legged Barsoomian thoats thundering across Martian dunes. But life on the level of microorganisms, living in sheltered environments, that's another matter. If it was there once, it could still be there now. To find it will take more than robotic probes with limited mobility, dexterity, and perception.

The *Viking* orbiters and landers kept on with their science observations long after the biology experiments came to a close. Orbiter 2's last transmission came on July 25, 1978, followed by the demise of Lander 2 on April 11, 1980, nearly two years later. Orbiter 1 sent its last signal on August 17, 1980, while Lander 1 signed off on November 5, 1982.

The Soviet space program attempted two launches in 1988 to explore Mars and its moon Phobos. But they met with disappointment, continuing a streak of bad luck that has plagued every Soviet or Russian Mars mission. (Out of more than sixteen attempts, none were successful.) The United States' Mars program has also had to deal with failures. The *Mars Observer* spacecraft carried seven instruments intended to investigate Mars over the course of a Martian year. The mission would "rewrite the books" on Mars, or so researchers hoped. But just days before the spacecraft was due to enter orbit around the Red Planet, it fell silent. In attempting to reconstruct what may have happened, engineers have surmised that a fuel line ruptured as the spacecraft prepared to fire up its engines to slip into Mars's orbit. Whatever the cause, after a seventeen-year hiatus, America's exploration of Mars appeared headed for the deep freeze.

Mars Exploration Makes a Comeback

But NASA's new administrator, the eccentric but very creative Dan Goldin, had an idea. Instead of launching a single massive spacecraft to the Red Planet, Goldin's new "faster, better, cheaper" plan sent a series of small spacecraft to orbit and land on Mars. This program started in late 1996 with the launch of the *Mars Global Surveyor* (MGS) spacecraft and the *Mars Pathfinder* mission. About half the size of *Mars Observer*, the *Surveyor* began mapping the Red Planet from polar orbit in March 1999 and continued to do so successfully through 2007. Among its discoveries have been altimetry data revealing a large basin of depressed and relatively uncratered terrain in Mars's northern hemisphere that is flatter than anything on Earth except for the sea bottoms, indicating the previous existence of a northern ocean. Possibly even more exciting are a pair of photographs that MGS took of the same crater in 2001 and 2005, which show the appearance of a new water erosion gully during the period between them. This could only have been created by a transient outflow of water from the crater side sometime between 2001 and 2005, thus proving the existence of subsurface reservoirs of liquid water on Mars today. *Mars Pathfinder* landed on Mars on July 4, 1997, with the help of parachutes, braking rockets, and airbags. Surviving several forty- to sixty-miles-per-hour bounces along the surface, *Pathfinder* opened and released a tiny rover dubbed *Sojourner* (after antislavery heroine Sojourner Truth). *Sojourner* then traveled for two months about the Ares Vallis runoff channel landing site, collecting geological information, and making the notable water-indicative discoveries of rounded cobbles and conglomerate rocks.

While the US robotic Mars Exploration Program accelerated following its twin successes of 1996–97, budgetary difficulties and bad luck threw the Russian program into chaos. Russia's Mars 96 mission aimed to place a spacecraft in orbit around Mars, as well as two small science stations and two ground penetrators on the Martian surface, but it was thwarted by a launch vehicle failure in the fall of 1996. This caused the indefinite postponement of a second mission, Mars 98, which was to deliver an orbiter, rover, and balloon to the planet. The Russian *Marsokhod* rover would have dwarfed the American *Pathfinder* rover and, instead of venturing just ten meters away from its landing site, could have logged nearly fifty kilometers. Trailing an instrument-laden "snake," the balloon, a product of the French space agency CNES (Centre national d'études spatiales), was designed to soar as high as four kilometers into the Martian atmosphere during the day but settle toward the ground during the Martian night. Designed for a ten-sol flight, the balloon would have been able to rack up several thousand kilometers in its windborne wanderings across the Martian surface. As Russia's economy continued to lurch through the next decade, however, hope for this mission dimmed, then disappeared.

The US Mars Exploration Program also had its share of bad luck, losing both its *Mars Climate Orbiter* and *Mars Polar Lander* due to failed orbit capture and landing maneuvers, respectively, during the fall of 1999. NASA pushed on with its decadal plan, however, successfully putting the *Mars Odyssey* spacecraft into orbit around the Red Planet in October 2001. Still operational two decades later, *Mars Odyssey* has used its infrared camera system and gamma ray spectrometer to map the mineral content of the Martian surface, discovering, among other things, continent-sized, high-latitude regions where the soil is more than 60 percent water by weight.

Following up on this success, as part of the Mars Exploration Rover (MER) project led by Cornell geologist Steven Squyres, NASA aimed two medium-sized rovers at the Red Planet in mid-2003. Arriving at Mars six months later, both rovers survived airbag-cushioned landings that set them down at two widely separated spots on Mars. The rover *Spirit* landed January 3, 2004, on the floor of Gusev Crater, a

Connecticut-sized feature 15° south of Mars's equator. Gusev intrigued mission planners because of a meandering, nine-hundred-kilometer-long valley that cuts into the crater's rim. The valley appears to have been carved by flowing water in ages past, and water may have once filled the crater. *Spirit* would be looking for present evidence of a watery past within Gusev. As it turned out, *Spirit* would have to do some hard traveling from its landing spot to find that evidence. The second rover, *Opportunity*, landed three weeks later nearly halfway around the planet in the Meridiani Planum, one of the smoothest and flattest spots on the planet. Again, mission planners followed evidence of water in selecting the landing site. Instruments aboard *Mars Global Surveyor* indicated that the area was rich in the mineral gray hematite. This form of iron oxide is found on Earth and usually forms in wet environments. As *Opportunity* began returning its first images, the MER team was amazed and delighted to realize that the rover had landed, rolled into a small crater, and sat on the surface facing an outcrop of layered rock. Few could imagine a sweeter spot to land.

Both rovers were slated to undertake ninety-sol missions. Before their nominal mission time was up, both found hard evidence of liquid water in Mars's past. In March 2004, the MER team announced that *Opportunity* had discovered solid evidence for liquid water in both the composition and morphology of rocks the rover inspected. A few days later, the team reported on findings from *Spirit* that hinted at a watery past for Gusev Crater. With these successes under their respective belts, or motherboards, the two rovers really got down to work.

Over the course of the months and years that followed, the rovers continued science investigations, scuttling about the surface of Mars while surviving Martian winters, dust storms, mechanical failures, and transient bouts of "amnesia." *Spirit* lived up to its name, overcoming numerous difficulties including a front right wheel that began to fail on March 13, 2006. The finicky wheel prompted the rover's driving team to drive the rover backward, with the cranky wheel dragging along the surface during forays over smooth ground.

The rovers have dusted off rocks, ground the surfaces of boulders, examined soils in microscopic detail, witnessed transits of the moons of Phobos and Deimos, captured images of dust devils skipping across the surface of Mars, and even watched for meteors in Martian skies. The library of raw images from the rovers is vast, ranging from exquisite, panoramic landscapes to close-ups of tiny grains of sand. They're available for viewing over the internet, assuming you have time to flip through more than a quarter million snapshots.

In May 2009, *Spirit*'s wheels broke through a crusty surface layer and sank into loose sand. The sand eventually proved to be too strong a trap, and NASA declared the tough little rover a stationary science platform after it had logged 7,730 meters of travel (4.8 miles). Even so, researchers examined the soil *Spirit* had disturbed as it tried to free itself, and a detailed study of the exposed soil layers once again revealed evidence of water in Mars's past. Just short of a year later, the rover fell silent, its last transmission occurring in March 2010. Meanwhile, *Opportunity* kept plugging away, trekking an epic forty-five kilometers (twenty-eight miles) from one landmark to the next before finally failing in June 2018.

In 2004, the first European probe reached the Red Planet. An ambitious mission, *Mars Express* included both a French- and Italian-made orbiter and the British lander *Beagle II*. Although the *Beagle* crash-landed, the orbiter proved successful, returning mountains of data including the first detection of trace quantities of methane in the Martian atmosphere. While initially disputed, these measurements have since been conclusively confirmed. In 2009, a team of researchers based at the NASA Goddard Space Flight Center announced that they had not only confirmed the presence of methane in Mars's atmosphere from ground-based observations, but had discovered multiple "substantial plumes" of the gas that blossomed during the warmth of the spring and summer seasons. Located in the planet's northern hemisphere, the plumes were seen over areas that bear evidence of ancient ground ice or flowing water.

To add to the intrigue, the research team determined that there exists on Mars a mechanism that removes methane from the atmosphere at a rate much higher than can be explained by photochemical destruction by ultraviolet light alone. Rather than taking centuries, something destroys Mars's atmospheric methane in as little as four Earth years, perhaps even as quickly as just a half year. Thus, the presence of methane on Mars today demonstrates that *there must be a source creating methane on Mars today.* This can only be explained by biology or hydrothermal geology, meaning that there is either life, or, at minimum, a subsurface environment friendly to life. Thus, if human explorers were to drill down and sample such environments, the chances are excellent that they could find life. Examining such organisms in their laboratory, astronauts could then determine whether Martian life followed the same biochemistry exhibited by all life on Earth or developed along different lines altogether. All Earth life—whether mushrooms, people, crocodiles, or bacteria—exhibits identical biochemistry, incorporating the same amino acids, and using the same RNA/DNA method of transmitting structural information from one generation to the next. But does life have to operate that way? Is life as we know it on Earth the pattern for all life everywhere? Or are we just one peculiar example drawn from a much vaster tapestry of possibilities? These questions are key to our understanding of the nature of life itself. *Mars Express*'s methane discovery tells us that the answers may wait for us on the Red Planet.

NASA's *Mars Reconnaissance Orbiter* (MRO) pulled into orbit in 2006 and began mapping the planet with a camera good enough to see the *Spirit* and *Opportunity* rovers from space and guide them in their travels. With the aid of its photos, we can now select boulder-free landing sites to help ensure the safety of future explorers, both robotic and human. In 2008, the *Phoenix*, so named because it was built using the spare test unit left over from the failed 1999 *Mars Polar Lander* program, redeemed that lost cause by landing successfully on the north pole of Mars. Conceived and led by University of Arizona planetary scientist Peter Smith, the *Phoenix* mission found pure water ice, putting to rest the debate over the composition of the Red Planet's bright white polar caps that had raged ever since their discovery by Christiaan Huygens in 1672.

The next NASA mission to go to Mars was the Volkswagen-sized *Curiosity* rover, which landed on Mars in August 2012 and is still operational today. Powered by a radioisotope generator that gives it the ability to operate regardless of sunlight or season, *Curiosity* carried eleven times the payload of scientific instruments wielded by *Spirit* or *Opportunity* and can travel much farther and faster. A sense of the advancement (and investment) this mission represents may be gathered by the figures in Table 2.1, which compares *Curiosity* to its predecessor MER rovers.

TABLE 2.1 Comparison of *Curiosity* with MER's *Spirit* and *Opportunity*

Specification	MER	Curiosity
Vehicle Length	1.57 m	2.7 m
Vehicle Mass	174 kg	900 kg
Instrument Mass	6.8 kg	80 kg
Power Supply	Solar	Radioisotope
Average Power	24 W	125 W
Average Vehicle Speed	100 meters/day	300 meters/day
Flash Memory	256 MB	2 GB
Computer Speed	35 MIPS	400 MIPS
Landing System	Airbags	Rockets
Cost	$400 million (each)	$2,300 million

Curiosity carries a suite of cameras, allowing it to take three-dimensional, true color photographs and movies, developed by the imaging experts at Malin Space Science Systems with input from *Avatar* filmmaker James Cameron. Its robotic arm is equipped with a microscope powerful enough to see fossils of microorganisms should they exist in the rocks or soil it encounters. It is also armed with a Los Alamos lab laser capable of vaporizing rocks up to seven meters (nearly 30 feet) away, and a French spectrometer to analyze the chemical composition of the vapor thus produced. Additional instruments to determine the elemental composition and mineralogy of soil samples include a Canadian alpha particle X-ray spectrometer, a Russian neutron spectrometer, and an American X-ray diffraction/X-ray fluorescence instrument. *Curiosity* also carries a very advanced gas analyzer, jointly developed by NASA and the French space agency CNES, that is not only able to sniff for traces of organic gases such as methane in the Martian atmosphere, but—based on isotopic composition—can also distinguish whether such gases are of geochemical or biological origin. A Spanish meteorological package allows it to measure atmospheric humidity, pressure, wind velocity and direction, air and ground temperature, and ultraviolet radiation. Finally, *Curiosity* is equipped with an American/German instrument called RAD (Radiation Assessment Detector) that measures and characterizes the radiation spectrum on the Martian surface for the purpose of preparing the way for human explorers.

Curiosity was a high-risk mission, as NASA defied the many-small-probes-instead-of-one-big-one wisdom it learned after the *Mars Observer* failure. Indeed, the program had one near-death experience, even before launch, when in 2008 NASA's science chief got cold feet and attempted to pull the plug on the mission based on a projected 20 percent cost overrun (after the other 80 percent had already been spent). Only a sharp counterreaction by the mission's defenders (including this author) and the gutsy willingness of NASA's tough administrator Mike Griffin to take responsibility saved the day.

Even riskier, *Curiosity* employed a very novel landing system, differing completely from both the Moon-lander-style thruster system used on *Viking* and *Phoenix* or the airbag method employed by *Spirit* and *Opportunity*. Instead, *Curiosity* landed by hanging on a cable attached to a tractor rocket that lowered it to the surface like a load suspended from a crane.

So, things were tense at the Jet Propulsion Laboratory (JPL) on the night of August 6, 2012, when all communication from *Curiosity* was cut off for seven terrifying minutes following its carrier capsule's entry into Mars's atmosphere. Then telemetry arrived giving the news. The rover had landed. The mission team at JPL went nuts.

As of this writing, *Curiosity* is still going strong, providing visually stunning panoramas of Martian landscapes taken during its journeys, as well as a bonanza of scientific results, including direct detection of methane emissions from the Martian surface. Other probes reaching Mars during the decade that followed included the MAVEN orbiter (arrived September 22, 2014), which surveyed the dynamics of Mars's ionosphere; the European ExoMars Trace Gas Orbiter (arrived October 2016) whose mission is to assess the presence of aerosols, water vapor, methane, and carbon monoxide in Mars's atmosphere (ExoMars also carried Europe's *Schiaparelli* lander, which crashed); and the NASA *InSight* lander (landed November 26, 2018), which is using seismometers to study the deep interior and geophysics of Mars.

Meanwhile, the orbiters delivered the previous decade continued to make discoveries. In 2018, scientists using the MARSIS ground penetrating radar on the European *Mars Express* orbiter announced the discovery of an underground lake of liquid saltwater near the Martian south pole. That same year, the SHARAD ground penetrating radar team on NASA's *Mars Reconnaissance Orbiter* announced the discovery of massive ranges of glaciers on Mars, covered only by a few meters of dust, extending down from the poles to latitudes as far as 38 north (the same latitude as San Francisco), and containing an amount of water equal to 10 percent of the freshwater supply of Earth. Then, in the fall of 2020, MARSIS scientists announced the discovery of three more underground saltwater lakes on Mars. Based on such data, it is becoming increasingly likely that we will discover not only the remains, but even living survivors of ancient microbial life on the Red Planet. As humble as such Martian microbes might be, the implications drawn from their existence are spectacular: The processes that lead to the origin of life are not peculiar to the Earth.

If we combine such a finding with the Kepler space telescope mission's discovery that most stars have planets, and that virtually every star has a region surrounding it—near or far depending upon the brightness of the star—that can support the type of liquid-water environments that gave birth to life on Earth and Mars, the conclusion would have to be that a very large number of stars currently possess planets that have given rise to life.

The evidence that could prove this case may be waiting for us on Mars. The Martian surface today lacks liquid water and so is very unlikely to support life. But the planet once had rivers, lakes, and seas. If living organisms were ever there, they would have left fossils or other biomarkers behind. In 2020, NASA sent the *Perseverance* rover to Mars to hunt for them.

Perseverance did not go alone. Simultaneously with its arrival on Mars in February 2021, two other missions reached the Red Planet. These were the United Arab Emirates' *Hope* orbiter, whose mission is to study Mars weather, and the Chinese *Tianwen-1* (Heavenly Questions) orbiter and *Zhurong* lander. Named after a Chinese mythological figure associated with fire and light, the *Spirit*-sized, solar-powered *Zhurong* successfully landed in May 2021. These feats were outstanding accomplishments for their sponsors. The UAE's *Hope* was the first interplanetary mission performed by any small country, and really puts the question to the rest, from New Zealand to Chile, as to why they aren't taking part as well. China, on the other hand, pulled off in a single mission what had taken America four phases—from flyby to orbiter to lander to rover—to accomplish. Where the Soviets and Europeans had repeatedly failed, China had succeeded on its first try, becoming the only country other than the United States to land a rover on Mars.

The main event, however, was *Perseverance*. From the engineering point of view, "Percy" is a close copy of *Curiosity*, using the same landing system and surface car design. But Percy carries a new set of instruments and experiments. These include the MOXIE device for extracting oxygen from Martian carbon dioxide and the *Ingenuity* helicopter.

Designed by Michael Hecht of MIT based on earlier work by Robert Ash, JPL-scientist-turned-Old-Dominion-University-professor, MOXIE uses high temperature zirconia-based solid oxide cells to electrolyze carbon dioxide into carbon monoxide and O_2. In tests so far, it has operated flawlessly, demonstrating in actual use on Mars a valuable technology for making oxygen, a key propellant and life support consumable.

Ingenuity is a battery-powered electric helicopter using a single set of counter-rotating blades to fly, with a small set of solar panels lining its sides allowing it to recharge its batteries between flights. Starting with flights of just a few meters, *Ingenuity* has been lengthening its excursions, and is now flying hundreds of meters to perform reconnaissance scouting for the *Perseverance* rover. This is still small potatoes. *Ingenuity* is a little machine, the size of a toy helicopter, and only made it to Mars as a hitchhiker payload considered unessential for the mission. But the *Sojourner* rover that *Pathfinder* took to Mars in 1997 was a toy-sized hitchhiker, too. Now we are sending rovers to Mars the size of real cars. In a decade we may be sending flyers the size of real helicopters.

The advantage of such flying vehicles over rovers is as great as that of rovers over stationary landers. The mobility of the *Viking* lander was limited to the two-meter reach of its robotic arm. Ground rovers have expanded that greatly. But a *Spirit*-sized helicopter could travel not hundreds of meters per day, as the rovers can, but hundreds of kilometers. Furthermore, they would not be limited by terrain obstacles, being able to easily travel over boulder fields, across canyons, or even land in canyons to explore their floor regions, then ascend and fly out again. They could do aerial surveys, carrying ground-penetrating radar to search for subsurface water or caverns with mobility unmatched by rovers and resolution unmatched by orbiters. When human explorers reach Mars, helicopters could serve them as scouts, or even provide fast long-distance surface-to-surface transportation. The potential is limitless.

To the professional Mars scientific community, however, *Ingenuity* and MOXIE are just along for the ride. They are interested in what Percy can do in the way of rock collecting.

Perseverance carries an advanced microscope, called SHERLOC, which might enable it to identify microfossils found in rocks. But, as the *Viking* mission showed, there are limits to what you can conclude from the data provided by remote-operated instruments. Many scientists really want to get some rocks home, to Earth, so they can study them directly in their labs, using the full suite of analytical capabilities available on Earth. Therefore, a principal goal of *Perseverance* is to wander about identifying rocks of interest, collecting, and stashing small samples of them for later pickup and return to the home planet.

This concept, known as the Mars Sample Return (MSR) mission, is now the principal focus planned for the next decade by the international Mars science community associated with NASA and ESA

(European Space Agency). The plan is for NASA to land a Mars Ascent Vehicle (MAV) on the surface to receive samples either brought to it by *Perseverance* or to have its own rover bring to it the stashes Percy has left behind. The MAV would then take off and rendezvous in Mars orbit with an orbiter that ESA will provide, which will then take the samples from the MAV and return them to Earth.

I must say I have very mixed feelings about this plan. In the first place, it is much more complex, and thus riskier and more expensive, than it needs to be. The *Curiosity* landing system can deliver a thousand kilograms to the Martian surface. This is enough to deliver a MER-sized rover, plus a two-stage MAV capable of taking off from Mars and flying directly back to Earth without the need to perform an autonomous rendezvous and dock in Mars orbit with a hypothetical (and not yet funded) ESA Mars orbiter. The MAV will need to bring its own rover anyway, because there is no way one can guarantee in advance that Percy will be able to bring it the samples years from now when the MAV arrives on Mars. So, why make the rendezvous with Percy part of the plan? More important, putting a return orbiter as a critical link in the mission plan greatly multiplies mission risk, as it might fail to be funded, or successfully launched, or delivered to Mars, or perform the rendezvous, and so on. That part of the plan is only being done because "Planetary Protection" bureaucrats have dictated that the mission must "break the chain of contact" with the Martian surface, lest the mission inadvertently bring back the Red Death from the Red Planet. But this is crazy—there can't be pathogens on Mars because there is no macrofauna or macroflora there. Besides, more than five hundred kilograms of rocks ejected from Mars by meteoric impact land on Earth every year and have done so for the past three billion years. Studies of these rocks have shown that in the process of their ejection from Mars, flight through space, and landing on Earth, major portions of them were never raised above 40° C, meaning they were not sterilized. So, if we could get the Red Death from Mars, we already would have. Demanding heroic measures to quarantine a few kilograms of rocks brought back by the MSR mission makes about as much sense as spending billions to have the border patrol search all incoming cars for Canada geese.

But the real issue is that the MSR mission won't get to the heart of the matter. We know from fossil evidence dating back at least 3.5 billion years that life appeared on Earth virtually as soon as it could. This means that either life evolves quickly and spontaneously from chemistry, or that life is being spread in microbial form across interstellar space and readily takes hold as soon as it finds a habitable environment. Since the thickly CO_2-enshrouded early warm and wet Mars was very similar to the early Earth, it means the Red Planet almost certainly once had life, either seeded from interstellar sources, evolved indigenously, or transferred by meteoric impact from the Earth. It may well have gone extinct on the surface, but if it was ever there, it could still be surviving in liquid water reservoirs we now know still exist underground. We need to go to Mars, drill, bring up samples of subsurface water, and see what is there.

The key question is not whether there is life on Mars, but *what is its nature?* At the biochemical level, all life on Earth is the same. As discussed earlier, whether bacteria, mushrooms, grasshoppers, or people, all Earth life uses the same DNA/RNA genetic alphabet. That's because we all share a common ancestor. But what about the Martians? If we both came from a common source, our alphabets will resemble each other, as the English alphabet does that of French. But if each biosphere originated locally, they could be as different as English and Chinese.

The necessary program of drilling, sample taking, culturing, biochemical analysis, and related observations is far beyond the ability of robotic rovers. It will require human explorers on the surface of Mars to carry out such a quest. But it will be worth the cost and risk involved because it would not only once again astound the world with the daring, creative genius of freedom, but it would also provide answers to fundamental questions about the prevalence and potential diversity of life in the universe that thinking men and women have wondered about for thousands of years.

If we want to know, we need to go.

FOCUS SECTION

WHERE SHOULD WE
SETTLE ON MARS?

How, then, could Romulus with a more divine insight have made use of the advantages of a situation on the sea, while avoiding its disadvantages, than by placing his city on the banks of a river that flows throughout the year with an even current and empties into the sea through a wide mouth? Thus, the city could receive by sea the products it needed and dispose of its extra commodities. By the river the city could ring up from the sea the necessaries of a civilized life as well as bring them down from the interior. Accordingly, it seems to me that even then Romulus foresaw that this city would sometime be the seat and home of supreme dominion.

—CICERO, *On the Commonwealth*

Our program of robotic exploration of Mars has provided us with a map of a world with numerous possibilities for settlement. Where should we go?

In choosing the best locations for Martian settlement, the first consideration is Mars's fundamental asymmetry. Mars's poles tilt 25 degrees on its axis relative to the planet's orbital plane, an amount quite close to Earth's tilt of 23.5 degrees. This would suggest that Mars and Earth have similar seasons. However, while Earth's orbit around the Sun is nearly circular, Mars's is a somewhat eccentric ellipse, with the Red Planet coming within 207 million kilometers of the Sun at perihelion, but then retreating to 249 million kilometers at aphelion.

In contrast, Earth orbits at a nearly constant distance of 150 million kilometers from the sun. So, while the solar flux hitting Earth is steady all year round, on Mars it is *20 percent above average* when the Red Planet makes its closest approach to the Sun. This occurs during the summer in the southern hemisphere. The amount of sunlight then falls to *20 percent below average* during southern winter.

FIGURE 2.1. *Map of Mars. The southern hemisphere is all elevated terrain, except for the deep depression of the Hellas Crater. The northern hemisphere is generally low elevation, descending from the vast plains of Arcadia, Acidalia, and Utopia to the flat Vasitas Borealis, believed by many to be the former seabed of Mars's ancient northern ocean.*[1] *(Credit: NASA)*

As a result, seasons in Mars's southern hemisphere are much more severe than those in the northern hemisphere, where the eccentricity of the planet's orbit acts to counter the variation in solar flux that would otherwise result from the summer-to-winter transition. In the southern hemisphere, winters can get cold enough to freeze carbon dioxide out of the atmosphere, accumulating it on Mars's southern polar cap. Then in southern summer, southern solar flux increases by

44 percent, causing huge amounts of the frozen carbon dioxide to evaporate, increasing atmospheric pressure by as much as 40 percent and driving massive dust storms.

So, not to put too fine a point on the matter, the weather on Mars is better in the north.

There is another distinction between the northern and southern hemispheres of Mars: elevation. Except for the deep Hellas and Argyre Planitia basins, whose bottoms descend as much as eight kilometers below the "datum"—the arbitrarily chosen elevation that is agreed to represent the equivalent of Martian "sea level"—nearly all the southern hemisphere consists of highlands typically four kilometers or so above the datum. In contrast, except for the very high mountains of Tharsis, Olympus Mons, and Elysium Mons (which can rise as high as twenty kilometers above the datum) and highland areas in Arabia and Lunae Terra, the north consists mostly of lowlands three to five kilometers below the datum.

For most purposes, low altitude locations are the best for Martian settlement, because the air there is thicker, providing superior radiation protection. In addition, thicker air makes parachutes and other aerodynamic landing and flight systems work better and is easier to pump up to higher pressures for use in propellant production or other resource-acquisition systems. Higher atmospheric pressure also makes ice more stable against evaporation, contributing, alongside the north's more stable temperature range, to the formation of the vast glacier formations and ice sheets that greatly enrich the resource potential of the mid- and high-latitude regions of Mars's northern hemisphere.

So, except for the Hellas and Argyre regions, the north beats the south regarding elevation criteria as well.

Another issue of interest, however, is solar flux, which is vital for both plant growth and solar-power generation. This is maximum at Mars's equator, falling to 70 percent of equatorial levels at 45 degrees north, half maximum at 60 degrees north, and 25 percent maximum at 75 degrees north. The equator is also the optimum place to launch from on Mars, as a launch site there offers access to a maximum range of orbit inclinations and imposes minimum propulsion requirements to reach orbit. If a Phobos skyhook system is built, as we shall discuss in Chapter 5, only rockets launched from equatorial locations will be able to make good use of it.

Most of Mars's equatorial regions are at high elevation, but there are some exceptions, including the southernmost parts of Amazonis, Elysium, and Chryse Planitias, and the four-thousand-kilometer-long canyon of the Valles Marineris.

The canyons and other outflow channels on Mars offer another advantage. They represent places where varied minerals and other materials present in the highlands may have been brought to low elevation by water action. Concentrated mineral ores are frequently formed in hydrothermal environments, which undoubtedly existed in the past in Mars's volcanic regions, and possibly in those heated by meteor impact as well. The volcanoes are generally found in very high-elevation regions. But some drain into canyons, as Tharsis does into the Valles Marineris. This makes the latter a place of particular interest as a site for Mars settlement and scientific exploration.

There are thus several competing criteria for choosing the best place to settle, with the sunny equator competing against the water-rich north, and specific locations offering potentially advantageous access to mineral ores.

I lead a nonprofit advocacy organization known as the Mars Society (www.marssociety.org). In 2019 and 2020, we held contests for the design of one-thousand-person Mars bases[2] and one-million-person Mars City States,[3] respectively. Between the two contests, some 276 designs were submitted, collectively representing a diverse set of technical, economic, social, political, and aesthetic concepts, with many of the teams proposing specific locations for their proposed towns. In

Table 2.2, I list those locations chosen by one or more of the forty-two semifinalist, finalist, or winning design teams published in notes 2 and 3.

TABLE 2.2 Proposed Locations for Mars Cities

Location	Latitude	Longitude	Altitude	Cited Advantage
Korolev Crater	73 N	163 E	-5 km	Ice-filled crater, nearby ore
Milankovic Crater	55 N	210 E	-5 km	Ice, near Olympus, cliffs
Tempe Terra	47 N	274 E	-1 km	Ice, flow channels, volcanoes
Arcadia Planitia	45 N	150 E	-3 km	Ice, low elevation, flat
Utopia Planitia	45 N	110 E	-5 km	Ice, close to volcanoes
Deuteronilus Mensae	44 N	20 E	-1 km	Ice, caves, channels, geology
Phlegra Montes	41 N	165 E	0	Ice, diverse topography ores
Chryse Planitia	30 N	300 E	-1 km	Wet soil, thorium, ore, channels
Jezero Crater	18 N	78 E	0	Old lake, minerals, sediments
Tikhonravov Crater	13 N	36 E	-2 km	Wet soil, ores
Libya Montes	3 N	85 E	-1 km	Tropical, access to Elysium
Echus Chasma	0	280 E	0	Tropical, varied geology
Meridiani Planum	0	0	-1 km	Tropical, minerals
Endeavor Crater	3 S	356 E	-1.5 km	Tropical, altitude, minerals
Gale Crater	5 S	138 E	-3 km	Tropical, low latitude, former lake
Valles Marineris	10 S	285 E	-4 km	Tropical, wet soil, channels
Arsia Mons	12 S	240 E	+7 km	Tropical, volcanic, ore, lava tubes
Gusev Crater	14.5 S	175 E	-1 km	Tropical, low altitude
Dao Vallis	36 S	90 E	-5 km	Low altitude, ice
Hellas	45 S	85 E	-7 km	Low altitude, lava tubes

Choose your location!

3

HOW TO GET TO MARS

ON MAY 5, 2021, Elon Musk's SpaceX team fired a shot heard 'round the world when they launched a prototype of SpaceX's Starship launch vehicle to twelve kilometers altitude and then returned it to land softly at its Boca Chica, Texas, launch site.

My wife, Hope, and I traveled to Boca Chica in early 2020 to meet with Elon Musk. While we talked inside the SpaceX on-site headquarters, a mariachi band played outside, providing entertainment for

FIGURE 3.1. *May 5, 2021—After five failures, SpaceX successfully lands its Starship orbital transport at its launch site in Boca Chica, Texas. (Credit: SpaceX)*

long lines of people queued up to apply for multiple categories of jobs. Hundreds were being hired to build Starships. Now, just a little more than a year later, one had successfully taken flight.

Starship is unlike any rocket that has flown before. It's a methane/oxygen-driven, stainless-steel, two-stage-to-orbit rocket with a payload capacity nearly equal to the Saturn V booster that sent Apollo astronauts to the Moon. The Saturn V, however, was expendable, with each unit destroyed during a single use. Starship will be fully reusable, like an airliner, and therefore promises a radical reduction in payload-delivery costs.

At the time of our visit, Starship had yet to be demonstrated. Yet here was Musk, building not the first experimental ship to prove the concept but, as we witnessed touring the place the next day, a shipyard, and a fleet. Is he mad? According to conventional aerospace-industry thinking, he certainly is. But there is a method to his madness.

SpaceX had been founded by Musk in 2002 for no less a purpose than *enabling the human settlement of Mars.* Cynics may doubt this statement of intent, based on the usually solid assumption that people who start businesses are in it for the money. They would be wrong in this case. I know Musk, and I played a role during that period in helping to convince him to make Mars his calling. He already had all the money he could want. If he wanted more, he knew plenty of easier ways to get it than to start a rocket company. He was looking to do things of immortal importance. Colonizing Mars, solar energy, and electric cars made the cut.

If you could convince NASA that you could send astronaut explorers to Mars for, say, $300 million each, the sale would be immediate. But interplanetary colonists could not afford anything close to such a price. A middle-class American willing to cash in his Earthly chips to move might be able to raise $300,000 by selling his house, and a working stiff might muster a similar sum by mortgaging his labor (as was done in colonial America, typically negotiating seven years of work for room and board as an indentured servant in exchange for one-way transport, which, as I noted earlier, would translate into about $300,000 today). But achieving such a ticket price would require cutting launch and space transportation costs by at least three orders of magnitude compared to those prevailing as of this writing. The affordability can only be done by making space systems reusable. A Boeing 737 costs about $100 million and typically carries about a hundred passengers. If it were expended after one flight, the ticket price would have to be set greater than a million dollars. But because the plane is reusable, you can fly to almost anywhere for less than a thousand dollars. To make colonization possible, the same economies will have to be implemented.

Hence SpaceX's key objective was to make launch vehicles reusable. The company began to make substantial progress toward this goal when, in 2015, after five failures, it succeeded in landing the first stage of its *Falcon 9* medium-lift launch vehicle safely back on a landing pad close to its Cape Canaveral landing site, saving nine of the booster's ten engines for reuse. It followed this feat in 2018 by introducing its Falcon Heavy, which increased its payload from the *Falcon 9*'s twenty tons to sixty tons, with all three of the nine-engine takeoff stages reusable, thereby saving twenty-seven out of the twenty-eight engines used per launch. But these feats, which cut launch costs by a factor of five compared to the previous state of the art and won SpaceX 30 percent of the world launch market, still do not approach the metrics necessary for colonization. Therefore, in 2019, after several years of preliminary design, Musk began to cut metal on prototypes of an entirely new type of spaceflight system—the Starship.

Now Starship test units have begun to fly. Furthermore, true to the projection Musk had shared with us in 2020, prototypes of the vehicle are being turned out in large numbers. This mass production approach has not only made possible an extremely forceful development program—launching, crashing, fixing whatever goes wrong, then trying again—it is also intended to enable a new concept of space operations. Starship operations will not be supported by three or four flight vehicles, but by scores of them, and eventually hundreds. Whereas NASA's Space Shuttle operations were measured in units of flights per year, Starship ascents will be counted in rates of flights per week, or even per day.

This is critical. Even though the Shuttle was mostly reusable, its average flight rate of four per year meant that, with an annual program cost of $4 billion per year, the real cost of each Shuttle flight was $1 billion. A Starship "transorbital railroad" employing five thousand people would probably cost about $1 billion per year to operate, including salaries, parts, fuel, and overheads. If SpaceX can manage two hundred flights—which could be done with twenty Starships, each turned around to fly again every thirty-six days, that would work out to $5 million per flight—1/200 the cost of the Shuttle with five times its payload, for a thousandfold improvement overall. Such are the numbers Musk is aiming for.

These breakthroughs will open the way to Mars.

It needs to be understood that what Musk and his team are doing is not just creating a revolutionary launch system. They are proving a larger point: It is possible for a well-led, entrepreneurial team to do things that, previously, it was thought only major-power governments could do, and that such a team could do them in one-third the time, at one-tenth the cost, and even accomplish things that had been deemed impossible altogether. As a result, SpaceX has unleashed an entrepreneurial space race. Worldwide, groups of engineers are now finding investors willing to finance new launch companies, spacecraft companies, and space technology companies.

There are already at least five companies in China working on launch vehicles that are strikingly similar to the SpaceX *Falcon 9*. This is not surprising. The laws of physics are universal. As soon as anyone anywhere demonstrates that a new idea works and is profitable, engineers elsewhere will have no difficulty finding investors willing to pay to duplicate it. Musk is thus sure to face competition, both from copycats and other innovators—which is why he is working hard to introduce a new type of launch system that will make his own currently market-leading Falcon rockets obsolete. As a result, space launch costs are set to crash.

The cheaper a space launch becomes, the more spacecraft and space missions there will be. It's simply supply and demand. Through mass production, the more spacecraft there are, the cheaper spacecraft will become. Furthermore, as space launch and spacecraft become cheaper, spacecraft designers can become much less conservative. When launches to orbit cost $200 million each, few spacecraft designers are willing to take a chance on using a novel technology subsystem that might offer improved performance. But if the costs drop to $20 million each, it's another story. So, space technology is about to take off, too, enabling much more capable spacecraft. Put cheaper launch together with cheaper and more effective spacecraft, and suddenly all sorts of space business plans that couldn't work before become extremely attractive.

Existing forms of space business, such as communications and remote-sensing satellites, will become much more widespread and profitable, but the benefits of the spaceflight revolution will not be limited to merely expanding the sorts of activities we see today. Entirely new kinds of space businesses, ranging from orbital research labs, industrial parks, and space hotels, to intercontinental fast travel will become practical. These enterprises will provide the market for mass production of ever-improving versions of Starship-type flight systems.

Could this affordable space-travel evolution really enable the settlement of Mars? I believe it can. Let's look at the numbers.

The cost of spaceflight can be radically reduced by making space systems mass-production items. A medium-sized rocket engine such as the Aerojet Rocketdyne RL10 has a mass of 300 kilograms and sells for $20 million. This is about one thousand times the cost of a new car, even though the car is a much more massive and far more complex piece of hardware. There are two reasons for this discrepancy. One is that Aerojet Rocketdyne has jacked up the price. (When it was produced by Pratt & Whitney, an RL10 went for $2 million.) The other is that the car is a mass-production item, while the RL10 is not.

In 2020, there were about 103 orbital launches worldwide. This number grew to 132 in 2021, and 174 in 2022 (sixty of which were SpaceX Falcons).[1] As launch costs drop, that number will continue to increase radically, perhaps tenfold, but such expansion would still be insufficient to make space launch vehicles a true mass-production industry. However, the advent of reusable launch vehicles will open an entirely new type of market for spaceflight—fast intercontinental travel.

Think about it. For the past three thousand years, people have made a lot of money sailing Earth's oceans. Some have acquired wealth by extracting resources from the ocean, for example, by fishing. But a far bigger business has been exploiting the fact that the oceans are a low-drag medium connecting port cities all over the world for transport by ship. Space is a zero-drag medium connecting every point on Earth. By flying through space, an orbit-capable rocket can fly from anywhere to anywhere on Earth in less than an hour.

It might seem that the energy requirements for spaceflight would make the cost of such travel prohibitive. But this is not so. The Starship system employs methane/oxygen propellant, which costs about $150 per ton. Between its Super Heavy booster stage (which lifts the Starship off the ground and propels it to one-third orbital velocity, after which it returns to the launch site) and the Starship itself, about five thousand tons of propellant are needed to fly to orbit, or to travel between any two points on Earth. This works out to $750,000 worth of propellant per flight, or $3,000 per passenger, assuming 250 passengers (half that of a Boeing 747) are onboard. Airline ticket prices are about triple the plane's fuel costs. If we assume that a spaceline can operate with similar margins, that works out to a $9,000 one-way ticket from Los Angeles to Sydney (or anywhere else), or $18,000 round trip.

I admit that I have never bought a round-trip airline ticket for anything like $18,000. But, in fact, the price of a first-class round-trip ticket from Los Angeles to Sydney, as of this writing, is $25,292, and all those fat cats in the front section get for the extra price is a bigger seat, a tablecloth, and a few drinks. They still must spend eighteen hours cooped up in an airplane. If they were to take a Starship, they could arrive in Sydney in less than an hour and enjoy thirty minutes of zero gravity as well as a view of the glorious starry sky of space along the way.

There were 174 orbital launches worldwide in 2022. Within a few years, there may be a thousand. But there are hundreds of intercontinental flights every hour. It's true that Starship-type vehicles won't be able to compete for all these routes. So long as Starships continue to employ vertical takeoff, they will make a lot of noise when they launch, so they will typically have to depart from offshore platforms or other locations well away from cities, with passengers delivered to the pad by short-distance aircraft shuttles. Even so, relative to satellite launch, the potential market offered by intercontinental fast travel is enormous.

So, now, let's fly to the Red Planet.

Starships will be able to reach low Earth orbit (LEO) carrying about one hundred tons of payload, but no propellant. To proceed farther to Mars, they will need to be refueled on orbit. Orbital refueling of rockets has never been done. But Starship is being designed to have such capability. SpaceX plans to create tanker Starships that will each carry a hundred tons of propellant to Earth orbit and mate with the interplanetary craft rear to rear. Then the two will be rotated, providing artificial gravity that will drive the liquid content of the tanker craft to its outlet pipes, from which propellant can be pumped to fill up the tanks of the interplanetary vehicle. This plan appears to be quite credible, so much so that in April 2021 NASA awarded SpaceX a $2.9 billion contract to develop this system to enable Starship to deliver astronauts to the Moon for its Project Artemis Moon base program.

Starship Performance

The performance of a rocket engine is given by two figures: its thrust and specific impulse. The thrust, given in pounds-force (lbf), is how hard it can push. Its specific impulse (I_{sp}) is the number of seconds it can get a pound of thrust out of a pound of propellant. As a crude comparison with automobile engines, you can think of thrust as being like horsepower and I_{sp} being like mileage. The Raptor engines used by the Starship have a thrust of about 500,000 lbf and an I_{sp} of 370 seconds if used in space and 320 seconds at sea level on Earth (it's lower at sea level because the Earth's atmospheric pressure pushes back on the rocket exhaust, thereby reducing its performance).

The Starship with both its stages fully fueled has a takeoff mass of 5,700 metric tons, or 12,540,000 pounds. To fly, it must achieve a total thrust greater than its liftoff weight. So, it uses thirty-three Raptor engines in its lower Super Heavy stage to take off and get going. Its upper stage, the Starship proper, will only have a mass of six hundred metric tons when it takes off from Mars, so three Raptor engines is more than enough.

The exhaust velocity (V_e) of a rocket engine, in meters per second, is 9.8 times its specific impulse. So, the Raptor will have an exhaust velocity of 9.8 X 370 = 3,726 m/s. The amount of propellant any rocket requires to perform a given change in its velocity, or ΔV, is given by the famous rocket equation:

Mass Ratio = (Wet mass)/(Final mass) = exp($\Delta V/V_e$) (3.1)

This is why rocket engineers will kill to get a few extra seconds of specific impulse. Their spacecraft's mass ratio, that is, the ratio of its "wet mass" (the vehicle plus its propellant) to its dry mass (the vehicle without propellant), increases *exponentially* in proportion to the factor $\Delta V/V_e$. The implications of this equation for the Starship's cargo delivery capability are explained further in the focus section at the end of this chapter.

Once it reaches Earth orbit, the Starship will need to change its velocity by a ΔV of 5.0 km/s to get to Mars on a six-month trajectory, perform midcourse maneuvers, and land. If you plug that requirement into equation 3.1, the result is that its initial mass in Earth orbit, fully fueled, will need to be four times its final mass landed on Mars with its propellant exhausted. The Starship is projected to have a mass of a hundred tons. If it is carrying a hundred tons of payload, the total ship dry mass will be two hundred tons, and it will need six hundred tons of methane/oxygen propellant to fly one way from low Earth orbit (LEO) to Mars.

The Starship could get to Mars faster than six months. If it flew without cargo and completely topped off its tanks with a thousand tons of propellant on orbit, it would have a mass ratio of eleven and would be able to reach Mars in half the time. But immigrants would not want to travel this way.

For one, using ten tanker flights instead of six would greatly increase the cost of the flight. Flying without cargo would also increase the passengers' ticket price, since cargo delivery could potentially pay for most of the cost of the mission. With one-way fares in the six-figure range, few immigrants will prefer to quadruple their costs to save three months of transit time. What's the rush? They are going to Mars for the rest of their lives. The cheapest way to travel is to ride the freight.

Furthermore, the six-month trajectory is actually safer than the three-month trajectory, not only because the six-month ship can carry more supplies and equipment and thus a much more robust life support system, but because the six-month trajectory is what is known as a "free return" trajectory. If you leave Earth for Mars on the six-month trajectory and decide to abort the mission, your orbit will take you back to Earth's distance from the Sun exactly two years after you left. This will be very fortunate for you because Earth will be there to meet you when you arrive. On the other hand, if you leave for Mars on a faster trajectory and need to abort, your path will loop out into the asteroid belt, and it will take longer than two years to return to where Earth could be but isn't. That would be very unfortunate.

The obsession of some in the NASA Mars human exploration mission design community to find faster ways to get to the Red Planet is mistaken. It will always be safer for astronauts to fly to Mars on the six-month, free-return trajectory. Even if superior propulsion systems, like nuclear thermal rockets, become available, the right way to use them would be to increase the mission payload, not cut the transit.

So, the six-month voyage it is, taking six tanker flights, each lifting a hundred tons of propellant to refuel. If the Starship launch and each tanker flight costs the same as an intercontinental flight ($2.25 million), this would add up to a cost of about $16 million to send a Starship one way to the Red Planet.

Starship is being designed to transport a hundred passengers to Mars. (It can carry more people on an intercontinental trip because such flights are much shorter.) Divided among a hundred passengers, the $16 million one-way price works out to a fare of $160,000 each.

There remains, however, the little matter of sending the Starship back. Returning a Starship from the surface of Mars back to Earth will require six hundred tons of methane/oxygen propellant. This can be produced on the surface of the Red Planet from locally available carbon dioxide and water, using one megawatt of power for about a year.

At current American electricity prices, this power usage would cost about $1 million. That's not too bad, but Martian power costs could be much higher, and in addition there would be the cost of mining the water ice, gathering the carbon dioxide, operating the propellant-making plant, plus the rental of the Starship itself, which will be away from Earth for 2.5 years. There won't be a lot of people taking the Starship back to Earth, so the outbound emigrants may have to cover this cost as well. This might add something like 50 percent to the overall cost, setting the ticket price at $240,000 each.

This is still within our target fare of $300,000. But if the emigrants are sharp, they may be able to make out a lot better. Musk told me he anticipates turning out Starships for a cost of about $10 million each. If that is true, then either he or his Chinese competitors should be willing to sell ships new for $20 million, and thrifty emigrants should be able to pick theirs up on the used intercontinental Starship market for $5 million. (That's right: If there are reusable launch vehicles, there will be *used* launch vehicles—something that has never existed before—and dealers selling them to those who can't afford to buy new.) Split that up a hundred ways, and it only adds $50,000 each to

FIGURE 3.2 *SpaceX vision of Starships being used to support the development of an early Mars city. (Credit: SpaceX)*

the basic $160,000 price. Moreover, when the new homesteaders arrive on Mars, the ship will be theirs to keep as starter housing, to peddle for parts, or to sell or use for profitable enterprises among the asteroids.

A Starship on Mars will be worth a lot more than a Starship on Earth. By delivering theirs to the Red Planet, enterprising emigrants will be able to make money off their trip. They could make even more if they take advance orders from Martians for terrestrial items in high demand and choose their ship's hundred tons of cargo accordingly.

So, pack your ship with stuff that's sure to sell for a nice markup, and ride the freight. You'll receive a great welcome when you arrive and do well by doing good.

FOCUS SECTION

CARGO DELIVERY CAPABILITY OF THE STARSHIP

Martian settlers will need to ship a lot of cargo to the Red Planet. How much freight can a Starship deliver?

A Starship has a mass of a hundred metric tons and a propellant capacity of a thousand metric tons. So, if it were not carrying any freight, it would have a mass ratio (wet mass / dry mass) of eleven. Its methane/oxygen propulsion system can deliver an exhaust velocity of 3.62 km/s. If you plug all that into equation 3.1, the result you get is a terrific ΔV capability of 8.7 km/s. But that is with zero cargo. If you load cargo, the ship's dry mass increases. The mass ratio then goes down, and so does the achievable ΔV. The results are shown in Figure 3.3.

The data in Figure 3.3 is quite interesting. It shows that a Starship carrying a hundred tons of cargo can still do 6.5 km/s, which is enough to take off from Mars and fly directly back to Earth. With two hundred tons of cargo, it can do 5.2 km/s, which is enough to fly from Earth to Mars on a six-month trajectory, perform necessary midcourse maneuvers, and land. It is also enough to fly directly from low Mars orbit to the main asteroid belt.

In contrast, the ΔV from low Earth orbit to the asteroid belt is 9.6 km/s, so even with zero cargo load, Starships cannot make that trip. The fact that Starships could readily reach the asteroid belt from Mars

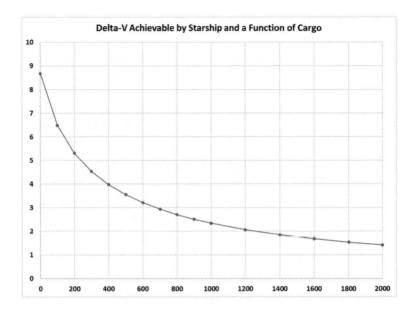

FIGURE 3.3. *ΔV capability of Starship in km/s (left axis) as a function of cargo load in metric tons (horizontal axis).*

but cannot do so from Earth has important implications for the economics of solar system development, which we shall explore further in Chapter 5.

With three hundred tons of cargo, a Starship can fly from low Earth orbit to Mars on a minimum-energy, 8.5-month trajectory, suitable for cargo-only deliveries (or really cheap ride-the-freight economy flights for immigrants on a budget). With four hundred tons of cargo, it can do 4 km/s, allowing it to deliver propellant from the Martian surface to orbit to refuel an asteroid miner Starship parked there.

Taken together, these capabilities would enable Starships to not only support the settlement of Mars, but to serve as vehicles for asteroid mining operations based on the Red Planet.

Finally, we note that even with two thousand tons of cargo, the Starship can still do 1.4 km/s. This is more than enough to allow a

Starship to launch from Mars and deliver a cargo to a tether skyhook system hanging down from Phobos and then return to land on the Martian surface. We will talk more about how such a "transorbital railroad" system could work in Chapter 5. Suffice it to say that once it is in place the export lift capability of Mars-based Starships will be truly enormous.

4

CREATING RESOURCES
ON MARS

THE GARDEN OF EDEN IS A MYTH. The first humans did not encounter bountiful Earth. Rather, they encountered a planet that was thoroughly unsafe, nearly uninhabitable.

In addition to being infested with voracious predators and deadly diseases, most of the Earth was far too cold for humans in the state of nature attempting to emerge from their tropical natural habitat in Kenya's Great Rift Valley.

Furthermore, there was almost nothing to eat. Berries were available, but only seasonally and in limited amounts, and if you ate the wrong ones, they could kill you. Roots could be found, but until fire was mastered, most roots were nearly indigestible. Mammals, birds, and fish abounded, but until hunting and fishing technologies were invented, they were extremely difficult to catch. There was nearly unlimited land, to be sure, but people cannot eat dirt or most of the things that naturally grow in it. Land did not become a source of food until people invented agriculture.

. . .

How Earth's Resources Were Created

Agriculture is a technology based on science. By science, I mean systematic knowledge of natural phenomena. In the case of agriculture, that knowledge is the understanding that plants grow from seeds, and that specific types of plants grow only from the seeds of that same type of plant. This is a fundamental regularity in nature, and once it was understood, humans could vastly expand their food supplies by creating farms.

But farmland is not a natural resource in any meaningful sense. Nature does not offer up land for our use. No fertile places exist in nature that are not already spoken for and occupied by species that are prepared to not only defend such bits of land to their deaths, but to immediately and vigorously counterattack and seize them back given the briefest opportunity. Farmland is a human creation, and its creation requires tools.

Humanity owes its power over the Earth to its virtuosity in making tools. Let us consider metal, the premier material for tools' manufacture, and thus for most of history held alongside land as one of the two primary natural resources worth going to war for. In Table 4.1, we list the natural abundance of civilization's most important metals, in parts per million of the Earth's crust, along with the approximate dates that each of them came into widespread use.

TABLE 4.1 The Abundance and First Utilization of Metals

Metal	Abundance (ppm)	First General Use
Copper	60	6000 BC
Silver	0.075	5000 BC
Gold	0.004	4000 BC
Tin	2.3	3000 BC
Lead	14	2000 BC
Iron	56,300	1200 BC

Metal	Abundance (ppm)	First General Use
Aluminum	82,300	AD 1890
Titanium	5,650	AD 1960
Silicon	282,200	AD 1960

The information in Table 4.1 should put to rest the idea that it is the Earth that provides our resources. For five millennia, stretching from the dawn of civilization until the beginning of the first millennia BC, human use of metals was limited to a list collectively comprising less than eighty parts per million of the Earth's crust. This was because of the limitations of human technology, in particular the temperatures attainable by early kilns. But when people developed the know-how for smelting iron, our metal resource base was expanded nearly a thousandfold. This development transformed human prospects by making metals cheap enough to be used in common tools, instead of being restricted to aristocratic weapons and ornamental artwork.

Among the tools eagerly put to use by the common folk were iron axes and iron-tipped ploughs. These redefined what land could be farmed, radically expanding agriculture, population, and urbanization, which in turn greatly accelerated the rate of innovation of society overall. Years later, with the development of electricity and scientific chemistry in the nineteenth and twentieth centuries, new metals, including aluminum, titanium, and silicon, became available. They may fairly be called new metals, because even though they were always here and present everywhere in abundances collectively exceeding even iron many times over, they were unknown. For thousands of years, migrating tribes, Roman legions, and wagon trains of westward-yearning pioneers had trod on trillions of tons of the stuff, but not one person among them ever knew these immense resources existed. *Because they didn't.* They did not exist until the knowledge to create them was developed. They are creations, whose genesis is human ingenuity—resources generated by human resourcefulness.

Aluminum foil and aluminum cans have replaced tin foil and tin cans, but the significance of these new materials reaches far beyond such beneficial substitutions. The new metals have enabled a new technological world virtually magical in its capabilities compared to the culture that preceded it, featuring air travel, spaceflight, computers, global telecommunications, pocket libraries, the internet, and perhaps someday, vast amounts of inexhaustible solar energy.

Any tally of a nation's "natural resources" today would certainly include its oil and gas reserves. But this was not always so. No eighteenth-century grand strategist would have given them a moment's thought. Petroleum was not originally a resource. It was always present where it can currently be found, but it was not useful. For most of human history, the vast majority of petroleum was undetectable and unobtainable, and the small portion that occasionally came into view by seeping to the surface was generally considered little more than a foul-smelling nuisance that ruined good cropland or pasture. It was only in 1859 when Colonel Drake drilled the first well in Pennsylvania that oil became a resource. Marketed as "rock oil," a cheaper substitute for whale oil, "petroleum" made nighttime lighting at home much more affordable (in inflation-adjusted prices, whale oil sold in 1860 for about $150/gallon), saving several species of whales from extinction in the process. A few decades later, with the invention of internal combustion and diesel engines, oil became the fuel for a radically expanded global transportation system, propelling vast numbers of cars, trucks, farm machines, trains, ships, and aircraft. It also became the foundation for gigantic industries producing plastics and synthetic fabrics, making clothes, and shoes, containers, and numerous other necessities of daily life far cheaper and plentiful than had previously been conceivable. As if that were not enough, drilling's associated product, "natural" gas, provided city lighting and then clean fuel for home cooking and heating, centrally generated electric power, and for any

number of industrial processes, improving public health while saving much of the world from deforestation.

All this from a resource that did not exist two hundred years ago!

There are things around today that clearly could become enormous
resources. Uranium and thorium were not resources until we invented
nuclear power. They are not yet major resources, because real or perceived problems have limited their application. Similarly, solar energy
is still a boutique energy source, because it is still too expensive, and
there are issues with storing it cheaply on a large scale to deal with its
day-night variation. These obstacles, however, are not fundamental.
Rather, they are fodder for inventors, and they can and will be solved.
Even more titanic energy sources, just over the horizon, await genesis
from our genius. These include fusion power, which will turn common light elements into fuel for artificial suns, providing the capability to transmute common raw materials and waste into unlimited
quantities of anything we need.

There is no such thing as a natural resource. There are only natural
raw materials. It is human ingenuity, manifested as technology, that
transforms raw materials into resources.

This view of the nature of resources—as human creations—is in
direct opposition to views that the environment contains a finite
amount of natural resources whose exhaustion entails our doom, or
alternatively, that the Earth has a finite "carrying capacity" sufficient
to provide the resources to sustain a limited number of people, but no
more. Such views are extremely dangerous, because they portray
humanity as fundamentally a race of parasites, at war with one another
for places in a shrinking lifeboat. While this worldview may seem
plausible, the above examples show, it is entirely false. This point is
critical, so let's really hammer it home.

Consider the territory currently occupied by the United States of
America. In the year 1500, it had a population of about three million people who possessed an advanced Neolithic culture. As a result

of their technology, including complex languages, leather clothing, housing, efficient control of fire, weapons that could kill at a distance, fishing gear, stone tools, canoes, pottery, domesticated dogs, limited agriculture, and extensive herbal, wildlife, and weather lore, those people were able to sustain themselves in a sometimes brutally cold environment that would have been uninhabitable for their distant ancestors in Kenya's Great Rift Valley. That said, they possessed almost no farmland, no metals except for insignificant amounts of natural copper, very limited textiles, no glass, and zero fossil fuel or uranium resources. Now consider the same region today, with its 330 million inhabitants, vast super-productive farmlands, and unlimited resources of iron, steel, aluminum, glass, plastics, fabrics, coal, water power, wind power, oil, and gas, readily capable of feeding and supplying itself with all necessities, as well as billions more.

It is the same place, but where it once had no farmland to speak of, it now has hundreds of millions of acres of the very best. That farmland was created by the pioneers and those who followed them, who cleared it, drained it, irrigated it, and did whatever else it took, including building roads and railroads to access it and make it ever more useful. It once possessed no iron, steel, oil, gas, or uranium resources. Those have been created, too.

So, what is it that "carries" us? It is not the Earth. It is human ingenuity.

We are the authors of the resources that now sustain us. We will create much more.

We created the resources of the Earth. We can create the resources of Mars.

How Resources Will Be Created on Mars

Resource creation on Mars will occur in three phases, dictated by the progressively expanding needs and capabilities of humans on the Red Planet. These phases are exploration, base building, and settlement.

The exploration phase will center on field research to resolve central scientific questions, such as those relating to the possible origin, extent, and diversity of life on Mars, and to assess the locations and amounts of potentially useful materials to support further development. This phase will involve at most a few dozen people on the Red Planet at any one time. The logistics and capabilities of ongoing exploration missions will be radically enhanced by the ability to transform locally available materials into propellants and life support consumables. Every kilogram of propellant that can be produced on Mars translates into one less kilogram of propellant that needs to be brought from Earth, which means one more kilogram of payload capacity to transport useful exploration equipment such as instruments or vehicles. Furthermore, once explorers can make propellants on Mars, they can use that capability to enable powerful long-range mobility systems—and mobility is the key to effective exploration.

Mars, however, has a surface area equal to all the continents of the Earth put together, and it contains varied geology as complex as any found on Earth. It is not a rock. It is a world. We will not be able to explore it with a few dozen people. Supporting larger numbers will require developing a much higher degree of self-sufficiency if logistical requirements are to be kept within reasonable bounds. The development of such technologies will therefore be a central priority of the first permanent human Mars base. As the amount of material that needs to be transported from Earth to support each person on Mars decreases, it will become practical to expand the base to hundreds and then thousands of people. As its population grows, the base will become an ever-stronger engine for invention and development. It will move beyond food and propellants to develop technologies to create structural materials, including brick, concrete, ceramics, iron, aluminum, glass, fabrics, and plastics. With these in hand, the base can begin manufacturing and building habitable and farmable space.

As these capabilities grow exponentially, the settlement phase will begin. Bases will grow into towns and then cities—the seeds of new branches of human civilization on the Red Planet. This phase will involve tens of thousands to millions of people on Mars who will need to create resources of every type, including an increasingly full spectrum of large-scale manufacturing and agricultural capabilities, and large-scale energy production to power it all.

Let's consider each phase of resource creation in turn.

The Exploration Phase

The essential resource creation capability required to support a robust and effective exploration program is the ability to produce propellants. The critical materials necessary for this activity are carbon dioxide and water, both of which are abundant on Mars.

Carbon dioxide comprises 95 percent of the Martian atmosphere and can be obtained anywhere on the planet. Back in 1976, in the very first paper ever written about Mars resource utilization, Robert Ash and his collaborators at Jet Propulsion Lab proposed acquiring CO_2 by freezing it out of the air,[1, 2, 3] using a windshield-wiper-type arrangement to stop the resultant dry ice from blocking the refrigeration surface. An alternative method of carbon dioxide acquisition with lower power requirements is to soak it up at low pressure with solid or liquid sorbents at cold temperatures, then release it at high pressure with heat. Starting in 1993 at Martin Marietta and then continuing from the late 1990s through 2021 at Pioneer Astronautics, I have led teams that have demonstrated both solid and liquid sorption pump techniques.[4, 5, 6, 7] They do indeed reduce power requirements. In fact, the natural day-night temperature swing on Mars is so large that it is possible to employ sorption pumps on Mars to acquire carbon dioxide from the air using no power at all! The sorption beds required by such systems would be rather massive, however, as they could only perform the temperature swing required to pump once every 24.6-hour Martian day.

But if you've got a little extra power on hand, the simplest way to get all the carbon dioxide you need is to run a roughing pump, compressing Martian air from 8 mb to 6 bars, at which point the CO_2 will liquefy at ambient Mars temperatures of -50° C. The argon and nitrogen components of Mars air won't liquefy at such temperatures, and so will be rejected (or kept separately for other uses, such as life support buffer gas) leaving you with pure liquid carbon dioxide that can be stored in a tank for future use as feedstock for propellant manufacture.

The only other ingredient you need to make propellant on Mars is water. There are three ways to get it: You can take it out of the air, bake it out of the soil, or melt it out of the ice.

Taking it out of the air seems simple. Air is everywhere. Unfortunately, Martian air only contains about 0.001 grams of water vapor per cubic meter. So, you need a very big fan using an awful lot of power to process a significant amount of water out of the atmosphere. Professor Adam Bruckner and his students at the University of Washington did the math in the early 1990s.[8] Their system, known as a WAVAR (Water Vapor Adsorption Reactor), employed a forty-kilowatt fan to drive very large volumes of Martian air through a sorbent bed to extract about forty-five kilograms of water per day. That's close to 24 kWh/kg of water produced, which is not very attractive.

In 2002, NASA's Mars *Odyssey* spacecraft reached Mars and mapped the water content of Martian soil from orbit. *Odyssey* found continent-sized regions at high latitudes with up to 60 percent water weight in the soil, with soil water concentrations of 5 percent being common at the equator. These results suggest that baking water out of the soil would be a much more efficient method of obtaining water on Mars than trying to take it from the atmosphere. This was demonstrated by Mark Berggren, Heather Rose, and the Pioneer Astronautics team in 2017.[9] In the Mars Water Acquisition System (MWAS) demonstrated by Berggren and Rose, Mars regolith simulant containing 5 percent water by weight was placed in a sealed vessel through which hot carbon dioxide was driven by a compressor. Water vaporized in the vessel was carried out with the CO_2 to be collected in a condenser vessel. A heat exchanger was placed between the input and output to the condenser to improve efficiency. By taking the water out of the soil as vapor rather than melting it out, pure water was obtained without saline contamination, a result that Rose dramatized by using the water product to both grow plants and as feedstock in a solid polymer electrolyzer.

The MWAS worked well enough, but while power requirements were an order of magnitude lower than that needed for a WAVAR-type system, they were still much higher than the 0.75 kWh/kg needed to vaporize water, let alone the 0.14 kWh/kg necessary to melt -50° C ice. In the 1600s, Huygens observed white caps covering Mars's poles and claimed that they represented frozen water ice. While endlessly disputed, the evidence supporting this hypothesis increased over the following centuries, particularly after 1971, when NASA's *Mariner 9* orbiter imaged networks of water erosion features across the Red Planet. All doubt was finally put to rest when Mars's north pole was definitively proved to be water ice by NASA's *Phoenix* probe, which landed there in 2007.

In 2018, the team operating the SHARAD ground-penetrating radar on NASA's *Mars Reconnaissance Orbiter* (MRO) reported the detection of massive formations of glaciers composed of pure water ice, covered by only a few meters of dust, extending from the north pole southward to 38 degrees N. This is the same latitude as San Francisco or Athens on Earth. The total amount of water in these glaciers is estimated to exceed the amount present in America's Great Lakes.

If the landing site is chosen in a location near such a glacier, water can be accessed at very low energy cost, using a proven technology known as Rodriguez wells, or Rodwells. Rodwells, which have been used for years to provide water to scientific bases in Antarctica, work by pumping hot water down into ice to melt some of it, thereby returning a larger volume of cold water to the surface. Some of this water can be reheated and sent back down again, repeating the cycle. The pumping power required by Rodwells is insignificant, enabling the production of water at little more than the tangible heat required to raise it to 0° C, plus the heat of fusion. On Mars this is 0.14 kWh/kg, so using 1 kW of waste heat, more than 150 kilograms of liquid water could be acquired each day. The vapor pressure of water at 0° C is 6 mb, which is less than average, ambient, low-altitude pressure on Mars. If the water were acquired from a high-altitude location, where ambient pressure is less than 6 mb, some kind of choke would need to be applied to the well head to keep the melted water in the liquid phase.

In 2018, the ESA *Mars Express* orbiter MARSIS ground-penetrating radar team discovered subsurface liquid saline water on Mars. While of great interest to exobiology research, these reservoirs are located more than a kilometer underground, and so represent a less accessible potential source of water for a Mars base than the glaciers discussed above.

So, Rodwells and glaciers it is.

THE CHEMISTRY OF PROPELLANT MANUFACTURE ON MARS

Once you have acquired carbon dioxide and water, you can start making propellant.

The simplest technique to make rocket fuel and oxygen on Mars is to electrolyze water to produce hydrogen and oxygen and then react the hydrogen with the CO_2 that comprises 95 percent of the Martian air as follows.

$$CO_2 + 4H_2 => CH_4 + 2H_2O \quad (1)$$

Reaction 1 is known as the Sabatier reaction and has been widely performed by the chemical industry on Earth in large-scale, one-pass units since the 1890s. It is exothermic, releasing energy as it proceeds, and so no power is required to drive it. It also occurs rapidly and goes to completion when catalyzed by ruthenium on alumina pellets at 400° C. In 1993, I led a team at Martin Marietta Astronautics in Denver that demonstrated a compact Mars propellant production system that worked by uniting this reaction with a water electrolysis unit and recycle pump.

As mentioned in the previous chapter, the number of seconds a rocket engine can squeeze a pound of thrust out of a pound of propellant is known as its "specific impulse," or I_{sp}. Roughly equivalent to the miles per gallon rating of a car, a rocket's I_{sp} is the top factor that engineers use in assessing the merit of its propellant mixture.

The methane/oxygen combination produced by the Sabatier/electrolysis system is excellent rocket propellant, offering a specific impulse as high as 370 s. This is the second highest specific impulse available from any plausible chemical rocket propellant combination, being only exceeded by hydrogen/oxygen, which can deliver 450 s. (For comparison a modern kerosene/oxygen engine can deliver 340 s while the ethanol/oxygen engines employed by German V-2 rockets in World War II achieved about 220 s.) Methane/oxygen, however, is three times as dense as hydrogen/oxygen, and can be stored in compact isothermal tanks because liquid methane and oxygen can store at the same temperature. Moreover, it takes much less power to liquefy and store liquid methane at 95 K than liquid hydrogen at 20 K. It was for this reason that I chose methane/oxygen as the propellant to use in the Mars Direct mission plan in 1990,[10] and Elon Musk chose it for his Starship Mars mission architecture in 2016.[11] Methane/oxygen was first demonstrated for use in 20,000 lbf Pratt & Whitney RL10 engines in the 1980s. Since that time, higher thrust methane/oxygen engines have been demonstrated, notably including the 500,000 lbf SpaceX Raptor and the 550,000 lbf Blue Origin BE-4.

The water produced by the Sabatier reactor can either be consumed as such or electrolyzed to make oxygen for propellant or consumable purposes, and hydrogen, which is recycled back to the reactor to make more methane. In the 1990s, when I set forth the Mars Direct plan, the availability of Mars water was unclear. I therefore proposed to bring the required hydrogen for methane/oxygen manufacture from Earth. Using this plan, 95 percent of the mass of the resulting propellant would come from Mars, with only 5 percent imported from Earth. Such a ratio is termed a "leverage" of 20:1. This would be reduced to 18:1 since it is desirable to run the rocket engine with a slightly fuel-rich (Oxygen/Fuel) mixture ratio of 3.5:1, rather than the stoichiometric O/F ratio of 4:1. Despite the availability of water ice on Mars, such a strategy to produce propellant at very high leverage might still be considered for initial missions as it reduces local material acquisition requirements for propellant production to simply gathering carbon dioxide, which can be done anywhere. The required hydrogen can be transported to Mars inside the ascent vehicle lower stage tanks that will later contain methane. For example, assuming two methane and two oxygen tanks per stage, we could first empty one of the lower stage methane tanks containing hydrogen to produce methane and water, with the methane product going into the upper stage and the water going into a compact bladder. Once that is done, the first lower stage tank could receive the methane produced from the hydrogen stored in the second lower stage tank. Then, with the transported hydrogen all gone, the system could proceed to make the rest of the propellant using hydrogen produced on an ongoing basis from the stored water.

A methane/oxygen engine operating with a specific impulse of 370 s, using propellant produced at a leverage of 18:1, may be said to have an "effective specific impulse" of 370 X 18 = 6660 s. *This is as high as that offered by advanced electrically driven ion drive propulsion.* But ion drives cannot deliver the high thrust needed to take off from a planet. A methane/oxygen rocket can. Moreover, it generates its thrust using a lightweight chemical engine, without the need to lug around a massive power plant. What local propellant production allows one to do is to take the energy produced by a large nuclear- or solar-power system on a planetary surface and integrate it over time, storing the energy in portable chemical form. This is by far the most effective way to use electric power for propulsion in space.

Another system that has been demonstrated for Mars resource utilization is direct dissociation of CO_2 using solid oxide zirconia electrolysis cells. The reaction is:

$$CO_2 => CO + 1/2O_2 \quad (2)$$

Reaction 2 requires a lot of energy (i.e., is highly endothermic), and the use of a ceramic membrane system with high temperature seals operating above 1,000° C. Its feasibility was first demonstrated by Robert Ash at the Jet Propulsion Lab in the late 1970s, and the performance of such systems was later significantly improved by Kumar Ramohalli and KR Sridhar at the University of Arizona.[12] Its great advantage is that no cycling reagents are needed. Its disadvantage is that it requires a lot of power—about twice that of the Sabatier process to produce the same amount of propellant. In 2021 a small-scale (20 gm oxygen/hour production rate) version of such a system, called MOXIE, was flown to Mars on the *Perseverance* rover.

Developed by Michael Hecht's MIT team, MOXIE worked like a charm.[13] However, because it involves the use of multitudes of small ceramic tubes, each tube with its own high temperature seal, questions remain as to the potential of zirconia technology to scale up to meet the production requirements for human Mars expeditions.

But you can accomplish the same thing using the Reverse Water Gas Shift (RWGS).

$$CO_2 + H_2 => CO + H_2O \quad (3)$$

This reaction is mildly endothermic and has been known to chemistry since the nineteenth century. If you combine it with a water electrolysis loop to recycle the hydrogen, you can go on forever producing any amount of carbon monoxide and oxygen you like from Martian carbon dioxide. The reaction occurs rapidly at 400° C, but it's a bit tricky to drive because its equilibrium constant is low, which means that it does not ordinarily go to completion, and it is in competition with the Sabatier reaction (1), which does. However, working at Pioneer Astronautics in 1997, Brian Frankie, Tomoko Kito, and I demonstrated that it could be done with conversions approaching 100 percent if the unreacted gas was rapidly recycled through a water condenser and air separation membrane and the right catalyst (copper on alumina) was used to make sure that only the RWGS reaction happened.[14]

If you want to, you can use the products of the RWGS directly in a carbon monoxide/oxygen rocket. NASA's Diane Linne has demonstrated such systems. But, while this eliminates the need to find water to make methane, the maximum specific impulse you get from such a rocket is only around 280 s. Now that we know that water is plentiful on Mars, it's clear that, for rockets, methane/oxygen is the right way to go. As we shall see, however, there are other uses for carbon monoxide.

The Base Building Phase

The activities of this phase center on creating a base of hundreds, and then thousands, of people who will not only support greatly expanded exploration, but the development of an ever-expanding technological, industrial, and agricultural capability to enable human settlement. To reduce its logistic requirements, the base will need to produce all its bulk materials. These include brick, concrete, ceramics, iron, aluminum, glass, fabrics, and plastics. Once we have these in hand, we will be able to create large amounts of habitable and farmable space and launch manufacturing and agriculture on the Red Planet.

LOCAL RESOURCE CREATION FOR LIFE SUPPORT

The same chemical systems used to make propellants can also be used to produce unlimited amounts of oxygen for life support purposes. Water acquisition systems that meet propellant production requirements can also easily meet life support system needs, as these are typically one to two orders of magnitude less. If we have water and CO_2, we can grow plants, supplying food and fabric. We can also extract nitrogen and argon buffer gas for breathing systems directly from the atmosphere, as these gases comprise 2.7 percent and 1.6 percent of the air there, respectively. Even though these buffer gases total less than 5 percent of the atmosphere, the disparity between atmospheric acquisition on a scale to meet propellant production requirements and those of life support means that any system designed to acquire enough Mars air to produce propellant will also process more than enough gas to meet crew buffer gas requirements as well.

BUILDING GREENHOUSES

We will be able to grow food on Mars for experimental purposes in small greenhouses transported from Earth. However, if we want to establish an agricultural base of sufficient size to support human settlement, we will need to make large-scale greenhouses on the Red Planet. Growing plants requires a great deal of energy in the form of light. For example, a prime hectare of Iowa land produces about twelve metric tons of corn per year. If double-cropped (as it would be under greenhouse conditions) and transformed into an equivalent mass of diverse foods, this could support about forty people. At high noon, such land receives about ten megawatts of sunlight, which works out to 250 kilowatts (kW) per person, or a day-night average of about sixty kilowatts each. This is forty times higher than average per-person residential electric power consumption in the United States. If this light were supplied artificially using solar energy, at least six hectares of land would need to be completely covered with 20 percent efficient photovoltaic panels to illuminate each hectare of cropland. (Considering the need to space out the solar arrays, it would take more than twelve hectares of photovoltaics to illuminate one hectare of crops.) While conceivably a highly efficient greenhouse system might significantly improve on these results, it is apparent that Mars settlers will want to take advantage of natural sunlight to support plant growth. Therefore, Mars settlers will need the capability to produce transparent structural materials in bulk. Such materials could include plastics, glass, and ice.

PLASTICS PRODUCTION

If you mix carbon monoxide with hydrogen, you get "synthesis gas." This is an extremely useful substance for making all kinds of fuels and plastics. (They don't call it synthesis gas for nothing.) For example, you can react synthesis gas with itself at 225° C at high pressure to produce methanol. The methanol in turn can be used to produce dimethyl ether (DME) by running it over a gamma alumina catalyst at 400° C. Methanol can be used in flex fuel internal combustion automobile engines and in fuel cells. DME can be used as diesel fuel, and in fact offers superior performance to petroleum-derived diesel fuel, possessing a cetane rating of 65, compared to petroleum diesel's 48. But once you have DME, you can use it to make ethylene or propylene. These substances are the basis of the plastics industry, as they are the raw materials for production of polyethylene and polypropylene, the two most important plastics.

In addition to supplying a source of hard plastic for the manufacture of spare parts by casting or 3D printing, polypropylene is an attractive material for the manufacture of fabrics for clothing and is currently widely used for production of superior outdoor attire. Many other synthetic fabrics, including acrylic, polyester, spandex, nylon, and synthetic suede, can also be made using related chemistry. These will probably provide most of the material for clothing on Mars, as wool, silk, and cotton, while producible, will be in short supply. (Don't panic yet, fashionistas. Polar fleeces and many other cool clothes are actually made from polyester.) Mars settlers will be able to clean their clothes without using any water, detergent, or chemicals, simply by putting them outside, as the low-pressure environment will remove all moisture and oils and kill all bacteria. I have experimented with this cleaning technique using vacuum chambers in my lab, and it works. The clothes feel and smell totally clean, and in fact they are clean as determined by sensitive instrumentation. The only downside is that stains are not removed, so they don't look clean. One remedy for this would be to use camo coloration for clothes, as it does not show stains. I predict this will be the style. (Now you can panic.)

Ethylene and propylene can also be used to create poly(methyl methacrylate) or PMMA, an extremely useful greenhouse construction material, generally known by its trade name, Plexiglas. An endless variety of other plastics, including strong structural ones such as nylon or PVC, can be made by combining carbon-hydrogen-oxygen compounds with nitrogen or chlorine, both of which are available on Mars.

GLASS PRODUCTION

The manufacture of clear glass on Earth requires silica sand (SiO_2), sodium oxide (Na_2O) from soda ash, calcium oxide (CaO) from limestone/dolomite, as well as feldspar (containing Al_2O_3). All the elements in these minerals are available on Mars, but not necessarily the minerals themselves. NASA's *Spirit* and *Curiosity* rovers have both detected large silica deposits in Gusev and Gale Craters, respectively. In 2012, scientists using the CRISM (Compact Reconnaissance Imaging Spectrometer for Mars) instrument on the *Mars Reconnaissance Orbiter* (MRO) announced that they had detected quartz, feldspar, as well as amorphous silica deposits near Antoniadi Crater on Mars. NASA's *Curiosity* rover has found calcium-sulfate veins in rocks, apparently left behind by water action in Yellowknife Bay. Sodium salts have also been found in various evaporate deposits explored by the Mars rovers.

It is thus apparent that the raw materials for glass production exist on Mars. Novel approaches may need to be developed, however, to minimize the complexity of production using the mineral feedstocks available within reasonable transport range of the Mars base.

Glass is an extremely important material for modern civilization. It may be noted that while glass has been used on Earth since circa 3500 BC, techniques for making transparent glass were not discovered until circa AD 100, in Alexandria, and truly clear glass had to wait for its development in fifteenth-century Venice. If Mars glass is to be made fully transparent, the elimination of iron-oxide dust impurities from the feedstock is of particular concern.

It should be noted, however, that optical-quality glass is not required to make many important glass products, including fiberglass, an excellent material for constructing various types of structures.

ICE ARCHITECTURE

We don't ordinarily think of ice as a prime structural material. That is because, where most of us live, ice doesn't last long. The indigenous Inuit people have a different point of view. That's because, where they live, it does.

On Mars, ice can last as long as rock. But unlike rock, we can, if we wish, melt it easily and cast it, or 3D print it, into any shape we like. So, why not put it to good use?

The discovery of copious supplies of water ice in mid latitudes on Mars opens amazing possibilities for large-scale construction of greenhouses and habitats. Employing Rodwells to extract water from glaciers, water can be obtained at an energy cost of 0.14 kWh/kg. A 1 MW$_e$ (1 million watts of electric capacity) reactor operating at 25 percent efficiency would produce 4 MW (4 megawatts) of heat (3 MW of waste heat from the reactor itself plus 1 MW of waste heat from whatever it is powering). If stationed near a glacier, this would be sufficient to acquire 685 metric tons of water per day. While weak in tension, ice is strong in compression, and structures can be built using blocks of it or any other conventional technique that is used for bricks, concrete, or other compression-strong/tensile-weak materials. However, as ice can be readily liquefied, it should be possible to 3D print ice structures in ways that are not possible with bricks. While a primary objective of a Mars structure is to contain pressure, the tensile weakness of ice can be productively remedied by supporting it from below with a pressurized membrane, whose pressurization itself is assisted by the weight of the ice above it. It takes a layer of ice about thirty meters thick to weigh down with a pressure of one bar on Mars. So, a polyethylene dome pressurized to 50 mb—sufficient for a greenhouse—could support and have its pressure contained by a 1.5-meter-thick layer of ice. A still more interesting possibility might be to cover a polyethylene dome habitat pressurized to 300 mb—Skylab space station pressure—with eight meters of liquid water with a thin ice layer on top. The inhabitants of such a dome would enjoy ample shielding, protected from cosmic radiation by a transparent aquaculture greenhouse, growing abundant kelp, fish, and other seafood above their heads.

Mars colonists could also build ice structures by melting tunnels into glaciers. Alternatively, Martians could use waste heat from nuclear reactors to melt the contents of ice-filled craters from beneath their icy surfaces, thereby creating large, habitable aquatic environments. For example, a five-hundred-meter-diameter crater twenty meters deep could be entirely melted out below its surface by the 1 MW_e reactor described above in about sixteen years, creating a large ice-covered lake that could be farmed for fish and kelp, and provide a convenient location for well-shielded submarine human habitats.

The unique advantages and potential opportunities offered by ice construction on Mars are profound. As we shall see in Chapter 7, they also open very exciting architectural possibilities for the creation of Martian cities of surpassing beauty.

BRICKS AND CONCRETE

Mars is rich in claylike materials suitable for brick manufacture. As early as the 1980s, Bob Boyd working at Martin Marietta demonstrated the production of bricks and "duricrete" concrete from Mars simulant soil. Mars also possesses gypsum, a key ingredient for the manufacture of Portland cement. Consequently, it should be possible to manufacture these traditional building materials on Mars.

Sulfur is plentiful on Mars, making the production of sulfur-based concrete readily feasible. Sulfur concrete smells bad, however, so it will probably only be used for outdoor structures like roads and landing pads.

Metals Production

Mars has plenty of iron. Indeed, the solid iron oxide feedstock, Fe_2O_3, is so omnipresent on Mars that it gives the planet its red color, and thus, indirectly, its name.

To get metallic iron from this rust, all you need to do is react the iron oxide with carbon monoxide, as people have done on Earth since about 1500 BC. The only difference is that instead of getting the carbon monoxide from incomplete combustion of wood, charcoal, or coal, on Mars we would get it from either solid oxide electrolysis of carbon dioxide or the RWGS, as discussed above. Alternatively, we could electrolyze Martian water to produce hydrogen, and react that with the iron oxide to produce metallic iron and water.

However, unless nearly pure iron oxide feedstock is obtained, there will be a significant additional task required to separate the metallic iron product from surrounding unreduced oxides of silicon, aluminum, calcium, and so on. From the early Iron Age to the present, this has generally been done on Earth by melting the entire reduced iron/unreduced oxide mixture at temperatures above 1,500° C. Since the liquid iron is heavier than the slag, the two readily separate. We could do the same thing on Mars, but the technique is very energy intensive.

A more innovative approach to separate the iron is to produce iron carbonyl, as long advocated by the University of Arizona's professor John Lewis.[15]

For example, carbon monoxide produced by the RWGS can be combined with iron at 110° C and high pressure to produce iron carbonyl ($Fe(CO)_5$), which is a liquid at room temperature. Then iron carbonyl can be poured into a mold, and then heated to about 200° C, at which time it will decompose. Pure iron, very strong, will be left in the mold, while the carbon monoxide will be released, allowing it to be used again. Similar carbonyls can be formed between carbon monoxide and nickel, chromium, osmium, iridium, ruthenium, rhenium, cobalt, and tungsten. Each of these carbonyls decomposes under slightly different conditions, allowing a mixture of metal carbonyls to be separated into its pure components by successive decomposition, one metal at a time.

An additional advantage of this technique is the opportunity it offers to enable precision, low-temperature, metal casting. One can take the iron carbonyl, for example, and deposit the iron in layers by decomposing carbonyl vapor, allowing hollow objects of any complex shape desired to be made. For this reason, carbonyl manufacturing and casting will no doubt also find extensive use on Mars, and the asteroid worlds that were Lewis's primary target, as well.

If pure metals can be obtained, 3D printing technologies can be developed and employed, as the feedstock will have predictable qualities. This will offer enormous advantages as it will allow Mars base personnel to produce anything they can draw, including spare parts for machines, an essential capability to reduce the logistics requirements of a base of any size.

ALUMINUM PRODUCTION

Aluminum oxide, or alumina Al_2O_3, is common on Mars, comprising about 4 percent of the planet's surface by weight. However, the aluminum is very tightly bound to oxygen, and so directly reducing aluminum in a manner like that employed for iron production would require extremely high temperatures. It is for this reason that, while iron has been in common use for three thousand years, aluminum metal was unknown to science until the nineteenth century and did not come into general use until the twentieth century. The process involves dissolving the alumina in cryolite at 1,000° C using carbon electrodes, which are consumed in the process, while the cryolite is unharmed.

The required carbon can be produced on Mars either by reacting carbon monoxide with itself to produce carbon dioxide and carbon, or by pyrolyzing the methane product of the Sabatier reaction.

But producing aluminum comes with a high energy cost, about 20 kWh/kg, ten times the 2 kWh/kg required to produce an equal mass of steel. For this reason, until a Mars settlement becomes truly energy rich, iron and steel will be the primary metals produced for use on the Red Planet.

GRAPHITE PRODUCTION ON MARS

In recent years, graphite products have become important structural materials, increasingly replacing both steel and aluminum. For example, top quality bicycles, once constructed of steel, and then aluminum, are now being made from graphite-derived carbon-carbon. Other graphite products, such as graphene and carbon nanotubes, hold promise for revolutionary applications ranging from superconductors and photovoltaics to ultra-strong fibers. Indeed, if the development of high-quality steel provided the basis for the Industrial Revolution of the nineteenth century, and the advent of aluminum enabled the jet-age globalized civilization of the twentieth century, it has been suggested that graphite may provide the material basis of the twenty-first century and beyond. If so, the raw material to make it is available everywhere on Mars and is as free as air.

Reducing carbon dioxide to make graphite is energy intensive. But if oxygen is already being produced for propellant purposes, then 0.375 kg of graphite can be produced for every kilogram of oxygen produced, *at no extra energy cost whatsoever.*

Technologies for producing good structural materials out of graphite could be of great value for a developing Martian civilization. Therefore, the creation of Martian civilization could prove to be decisive for the development of graphite—with extraordinary benefits to human civilization on Earth as a result.

Is there a better way to dispose of carbon dioxide than using it to create unlimited amounts of a new lightweight structural material with miraculous properties ranging from super strength to superconductivity? I don't think so.

There was a time when farmers regarded oil seeping out of their land as a pollution problem. That seems bizarre to us today, because oil is a prime resource for our civilization. Today we are concerned that we are producing excessive carbon dioxide. The citizens of a future graphite age might well look back on us and see such concerns as equally incredible.

The Settlement Phase

While exploring and base-building on Mars can be supported by government, corporate, or nonprofit largesse, a Martian civilization of millions of people will require an economic foundation. For the same reason that no nation on Earth is truly autarchic (and those that try to be, such as North Korea, are very poor in consequence), it will never be possible, or desirable, for Mars to be totally self-sufficient. That said, both the purchase and the transport of goods from Earth will need to be paid for, with the latter representing a formidable burden even if interplanetary transport costs can be brought down several orders of magnitude.

There are several ways that Mars can generate the cash it needs for its imports, and we will discuss them in the next chapter. But given their costs, it will remain imperative that everything heavy—for example, all goods too massive in respect to their value to use intercontinental air freight on Earth today—needed on Mars, be made on Mars. Only high-value, lightweight goods requiring Earth's vast division of labor for their manufacture should be imported from Earth. To achieve this objective, a vast range of technologies will be necessary, and the capabilities of the limited Martian workforce will need to be greatly multiplied in both quantity and diversity through broad application of labor-saving machinery, automation, robotics, biotechnology, and artificial intelligence.

The full array of local resource creation, manufacturing, and labor-amplifying technologies requiring development is too lengthy to be discussed in any detail here. However, energy is fundamental to the entire project. All the chemical elements we need are to be found on Mars. We know how to recombine them to create useful resources. But it will take plenty of energy to do that on the scale required for settlement.

Where will this energy come from?

Energy for Mars Settlement

On Earth, power can be generated by combustion of biomass or fossil fuels, tapping waterfalls and wind, converting sunlight by photovoltaic or concentrated thermal means, accessing geothermal sources, or via nuclear power. On Mars, combustion is not a net energy source, as both the fuel and the oxygen to burn it need to be made. Water and wind power are currently unavailable and will remain so until the planet is terraformed (more on that in Chapter 11). Solar energy can be used on Mars, and has been, but sunlight is only 40 percent as strong on the Red Planet as it is on Earth, and it is subject to monthslong cutoff by dust storms, making it even less attractive for large scale use. Geothermal power is a possibility, but only in limited amounts in limited locations.

The energy basis for a Martian civilization will need to be nuclear power.

Uranium and thorium have been detected on Mars widely dispersed at about 1 ppm (part per million) concentration, which is about the same as is typical on Earth. Concentrated uranium ores, typically 2 percent, but in a few cases up to 20 percent uranium, have been found on Earth, and there is no reason why they shouldn't exist on Mars as well. So, in principle, nuclear fission reactors could be fueled from uranium and/or thorium mined on Mars. The problem, however, is that refining high concentrations of these materials out of mined ores is a highly complex and involved industrial process, made much harder if it is necessary to enrich fissile material content from 0.7 percent U-235 to 3 percent U-235 via isotope separation to create reactor-grade fuel.

An initial solution might be to import uranium or thorium from Earth. If this is done, it would be extremely beneficial to use this material in breeder reactors. Existing nuclear power plants on Earth use only about 1 percent of the energy content of the mined uranium (i.e., the U-235 fraction, plus a little bit of U-238, which is transformed into plutonium as a by-product of normal reactor operations). They can get away with this because, even with low utilization, fuel costs are only about 5 percent of the cost of nuclear power. While indulging in such waste is tolerable on Earth, it would be very distasteful to Mars since (unless the fuel was highly enriched before shipping) it would multiply the mass of the fuel that needs to be transported across interplanetary space by two orders of magnitude. Thus, while breeder reactor technology has lagged on Earth, it is sure to be developed on Mars. The subsequent transmittal of this technology back to the home planet could cut the production of nuclear waste here a hundredfold.

There are two types of breeder reactors. One type uses a high energy ("fast") neutron spectrum to breed Pu-239 from U-238. This reaction was explored very early in the history of nuclear engineering because it was used to manufacture the fissile material for the Nagasaki bomb that ended World War II and for many other bombs since. The other breeding reaction uses "thermal" neutrons that have been slowed down by a moderator to breed Th-232 into U-233. This was explored by Oak Ridge National Laboratory's visionary director Alvin Weinberg in the 1960s but not developed further by the Atomic Energy

Commission or its Soviet counterparts because, unlike Pu-239, U-233 cannot be used to make bombs. However, as thorium is four times more common than uranium, reactors that can breed it into fissile U-233 represent a potentially superior energy technology. There are several entrepreneurial companies currently working on developing thorium breeder systems.[16] Such technology could be of great interest on Mars, and should the current terrestrial development efforts fall short, the Martians may well turn out to be the ones who take it across the finish line.

But Mars has an even better potential energy source: fusion power. Deuterium (^{2}H a.k.a. "D," the heavy isotope of hydrogen that can serve as fuel in a fusion reactor) is about five times more common in Martian water than it is on Earth (833 ppm on Mars versus 166 ppm on Earth). Because the atomic weight of deuterium is twice that of ordinary hydrogen, separating them is much easier than separating U-235 from U-238, which only differ in weight by 1.3 percent. Moreover, no mining is required, as a Mars colony will constantly be using water, and the physical chemical life support system will need to electrolyze at least one kilogram of water per day for every colonist. A Mars colony of one hundred thousand people will need to electrolyze at least a hundred metric tons of water per day, within which there will be 11.1 tons of hydrogen, from which 18.5 kilograms of deuterium can be extracted. This could be used as fuel in a deuterium fusion reactor, whose net reaction, including highly reactive side products tritium and helium-3, which will be consumed in situ, is:

$$3D => {}^4He + p + n + 21.6 \text{ MeV} \quad (4)$$

The primary energy cost of hydrogen distillation methods of deuterium production is electrolysis. In this case, the electrolysis of water for life support will require about 480 MWe/hrs, or 20 MWe all day long. The amount of deuterium, however, will be enough to produce

1850 GW-hours, roughly 3,850 times as much energy as it took to extract, or 1,500 times as much power if we assume the fusion energy is converted into electricity at 40 percent efficiency.

So, using 20 MWe for isotope separation, Martians can produce enough deuterium to generate 30,000 MWe of power!

If uranium can be mined and refined on Mars, the presence of plentiful deuterium would facilitate the building of reactors of the CANDU type, which use heavy water to moderate their neutron flux, and thus can employ natural uranium without isotope enrichment as their fuel. This would at least eliminate the need for the very difficult U-235/U-238 isotope separation from the fuel cycle. Alternatively, natural uranium or thorium could be put in a blanket surrounding a D-D fusion reactor, and making use of the reaction's neutron product, be bred into fissile Pu-239 and U-233, which could then enable uranium and thorium breeder reactors. Such a system would also serve to make importing uranium or thorium from Earth more practical, since no isotope separation would be needed on either Earth or Mars, and the useful energy content of the uranium per kilogram would be multiplied 140-fold by the fusion-driven transmutation.

But the simplest cycle is deuterium fusion, made very attractive by the abundant deuterium that will be produced as a by-product of the operation of Mars's settlement life support system. It is for this reason that, if fusion is not developed on Earth, it certainly will be on Mars.

Necessity is the mother of invention.

Utilization or Creation?

An environment becomes habitable once people develop the technology to create, out of local materials, the resources necessary to support human habitation. The NASA acronym ISRU, or "In-Situ Resource Utilization," is based on a fundamental misunderstanding of the nature of reality and the human condition. As humans have expanded our physical and technological reach, we have not "utilized" and thereby diminished a stockpile of "natural resources" drawn from some fixed inventory existing before us. Rather, we have *created* the resources available to support us on Earth, and continue to do so, with radically increasing effectiveness today.

There are no natural resources on Mars—or anywhere else in the universe—today. But there will be plenty of resources on the Red Planet once resourceful people arrive.

5

GETTING RICH ON MARS

THE FUNDAMENTAL REASON to colonize Mars is to establish new branches of human civilization. Through their creativity, these new societies will expand and enrich the human experience, generating new ideas, new literature, new discoveries, new inventions, new histories, and new heroes. That's what it is all about. The game of life is not played for cash. It is played for children, for posterity, for immortality. That is why humans need to go to Mars.

Our purpose is not to make money. Our purpose is to create.

It is not Greed that will take us to Mars. It is Hope.

In a fundamental sense, Plato was right: Ideas create reality.

Mars will not be settled by people who are looking for the easiest way to fill their bank accounts. Mars will be settled by people who passionately want to do something of transcendental significance and are willing to risk their lives and fortunes to do so.

In the beginning was the Word.

The settlement of Mars is being launched by an idea. That idea will require means for its realization. That takes money. But where there is a will, there is a way.

So, what's the way? How can the settlement of Mars be funded?

Initial funding of a Mars settlement could come from a government, or more likely, from a home front Mars settlement organization analogous to those that funded (and still help fund) Jewish emigration to Palestine before the establishment of the state of Israel, or which in earlier periods helped finance Mormon pioneers going to Utah or Pilgrims to Massachusetts. However, eventually, a settlement needs to start paying its way.

The cost of logistical support of a Mars colony can be greatly reduced by in-situ production of useful materials. Thus, there should be no need to import food, fuel, oxygen, fabrics, bricks, ceramics, steel, aluminum, plastics, glass, and all sorts of simple products made from them from Earth. But complex products require a large division of labor. Martians will certainly grow their own food using Martian carbon dioxide and water in greenhouses made of Martian glass or acrylic, irrigated using tubes made from Martian polyethylene. They may even make the motors and wires to drive the fans and pumps to ventilate and water the greenhouse. They can also make most of the heavy parts needed to build CNC (computer numerical control) machine tools. But what of the circuit boards and computers needed to control such systems? At a certain point, subsystems will be too complex to manufacture with the limited division of labor available on the Red Planet. Even a million-person Mars city-state will be extremely limited in its diversity of skills and productive capability compared to the many billions on Earth. Consequently, imports will be necessary. The cost of these imports will be greatly reduced by restricting them to low mass components and taking advantage of ever falling space transportation rates. But still, they will need to be paid for. How can Martians generate the cash?

Inventions

In my view, the best, early, large-scale source of cash income that Mars colonists can generate will come from the sale and licensing of intellectual property. This will come naturally from the nature of the Martians themselves and their situation.

The Martians will be a group of technically adept people in a frontier environment that will challenge them, indeed force them, to innovate. They will face a terrific labor shortage. This will compel them to innovate in the areas of labor-saving machinery, automation, robotics, and artificial intelligence. Limited to greenhouse agriculture, they will have a shortage of land and livestock. This will force them to innovate in the area of biotechnology, to create ultra-productive and highly nutritious crops. Lacking attractive sources of fossil fuels, wind or water power, or solar energy, they will be impelled to innovate in areas of nuclear power, including advanced fission designs like homogenous molten salt thorium breeder reactors (a.k.a. LIFTRs, for LIquid Fluoride Thorium Reactors), and fusion, as the deuterium fuel for fusion reactors is five times as common on the Red Planet as it is on Earth.

All these innovations will have tremendous utility on Earth. The Martians therefore will patent them and license the patents for use on the home planet. The revenue from such intellectual property sales could be enormous.

Indeed, were it necessary to propose the settlement of Mars to a group of investors as an attractive for-profit enterprise, the concept of creating an inventors colony would be an appealing way to go. Some might object that an inventors colony could much more easily be established on Earth. This is true. But a Martian inventors colony would have strong countervailing advantages.

First and foremost, the people would be much better. Those recruited for the Mars inventors colony would not be mere careerists looking for a comfortable, high-paying, or interesting job. They would be men and women with a cause. Nothing great has even been accomplished without passion. It's easy enough for an engineer looking for a job to tell a prospective employer that he or she is passionate about the work. But to pull up stakes and put your life on the line to open a new frontier on Mars, you really must have it. It's that kind of zeal that puts steel in someone's backbone, the same kind of steel that gave the colonists in New England, Utah, and Israel the fortitude they needed to overcome all odds. It's not a quality that can be bought.

Furthermore, it's a quality that circumstances would reinforce in a Mars inventors colony but subvert in one located in California, or anywhere else on Earth. Put simply, Earthlings can quit. Anyone who is part of a terrestrial inventors colony who is not happy with the way things are going, who loses interest, or who gets a better offer from somewhere else can just split. If you sign up for the Mars inventors colony, that's not true. You are committed, in it to win it. For you, it's Normandy Beach, do or die. You've got to make it work. So, you will.

If it's wise to invest in people who have skin in the game, the Martians will be very investible.

Napoleon said, "In war, morale is to material as ten is to one." If backed by similar funding, a terrestrial inventors colony could be much larger and better equipped than one on Mars. But when considering the odds of success in any venture, the quality of *resolution* is priceless.

Real Estate

Once a colony is established on Mars, it will have things besides intellectual property to sell. The most obvious of these is real estate.

The surface area of Mars is equal to all the continents of Earth put together. This plentitude of available land might seem to be a fatal disadvantage to would-be sellers, but that is not so. Land is only useful where the services necessary to its utilization are available. You can buy land very cheap in my home state of Colorado right now in places where there is no road access, electricity, or water. Some people actually do this. But far more prefer to pay orders of magnitude higher prices for land that is connected to the road and electric grid, and much more still if it has access to water and gas pipelines. While land exists irrespective of human actions, real estate is something that people create.

Land is not real estate. *Land is the raw material for the production of real estate.* Mars is extremely rich in this raw material. That makes it an attractive place to mass-produce real estate for sale. The equipment required to do this is the same as that needed on Earth: vehicle access, electric power, water, a supply line to food and other stores, and housing. Once it is established, a Mars inventors colony can provide all these things to neighboring land, thereby developing it into sellable real estate. Unless they are part of an organized group, willing and able to establish all these amenities in a totally new location, those who want to move to Mars will need to buy or rent such property.

This brings up a question. Why would anyone other than the first wave of zealot pioneers want to move to Mars? There are plenty of potential reasons. The first of these is freedom.

By freedom, I mean here the right to be what one wants to be and to live as one wishes to live. This can be either individual or collective in character. Both aspects played a prominent role among those who colonized America and other places in the past and are sure to play a similar role in the future.

To be free, it is sometimes necessary to go somewhere else and far away. This holds true for individuals being hunted by debt collectors, criminal gangs, political bands, or state authorities. It is true for those who would practice forbidden forms of enterprise, professions, or technology, who would give voice to forbidden ideas, or who find demands for conformity to new or old orthodoxies incompatible with their self-respect. It is also true for religious groups whose views and practices are deemed odious by surrounding society, or who themselves find life within a disrespectful surrounding culture intolerable. It is true for people seeking to implement novel political or social ideas. It is certainly true as well for people whose membership in certain racial, national, or ethnic groups marks them as targets for hate, or whose homeland has become inhospitable due to natural, economic, or human-caused disasters and are unwelcome elsewhere.

"The strong do as they will. The weak suffer what they must." So, according to the ancient Greek historian Thucydides, the powerful Athenians said when dictating surrender terms to the Melians in 416 BC. That was political reality then, and things haven't changed much over the twenty-four centuries since. Given human nature, it is likely to remain the case in the future.

We are all "the weak." Depending upon how social groupings are defined, everyone is a member of an outnumbered minority. We all face potentially overbearing majorities. Their freedom to make the rules "as they will" necessarily tramples that of those who would have things different.

The weak suffer what they *must*. But they need not suffer so much if they have someplace else to go.

Freedom is the ability to choose. That includes the ability to choose to live in a place where the rules are different, or better yet, haven't yet been written.

Mars could offer many such places. That gives it a great deal to sell.

But once things get going, there will be a new reason to go to Mars that will no doubt eventually draw many more than all the other rationales.

The pay there will be higher.

Precisely because of the severe labor shortage prevailing on Mars, wages will be high. This of course will drive Martian business toward labor-saving machinery. Any technology that multiplies the productivity of labor will be avidly sought, and precisely because the power of labor is multiplied, it will be possible to pay more for work. Furthermore, since businesses and city-states will be in competition for immigrant labor, they will need to treat it well.

It is this virtuous cycle of high-priced labor driving technology forward, thereby multiplying productivity, enabling still higher wages, that has drawn millions of immigrants to America from colonial times down to the present day. It will operate even more forcefully on Mars.

The value of any piece of Martian real estate is multiplied in accord with the services that are provided to it. These include power, water, ground access, food supply, and housing. So, it is not just the owners of the land itself who will profit from such sales, but all those who can provide these services, or the materials needed to provide such services. Thus, every branch of Martian industry and agriculture will get a piece of the action.

Invention and immigration will thus form two of the primary pillars of the Martian economy. Both are based on freedom.

Tourism

A Mars colony could also generate income by hosting tourists. These could include people visiting Mars for scientific, commercial, recreational, or medical reasons.

Scientific tourism will be a moneymaker for a Mars settlement right from the start. The world's space and scientific agencies are already spending billions of dollars to explore the Red Planet. A Mars colony could offer habitation, laboratories, vehicles, drilling rigs, robots, and all sorts of other vital equipment at a price that would be a bargain for NASA, ESA, and so on, but very lucrative for the settlers.

Businesses will also want to send representatives to Mars to make deals. A Martian inventors colony would be a technological gold mine, and astute businessmen will want to go there to see what it might have to offer. In addition, Mars itself offers commercial opportunities, including the development of mineral resources for export, as we shall discuss below.

Tourists will also want to visit Mars for recreational reasons. Such people will no doubt need to be well heeled because the trip is expensive and could require taking a couple of years off from income-generating pursuits. Yet there will be much to see on Mars, ranging from its natural wonders to the new world's novel architectural and social forms. So, some people will be willing and able to pay a hefty price for a first-class experience of such a visit. Some have argued that a Mars colony could be based on the business plan of catering to such tourists.[1] I don't think so. But if a Mars colony of one hundred thousand were visited by five hundred tourists per year willing to drop $1 million each to see remarkable sights and have a good time, that could provide 5 percent of the local population with $100,000 of income, or viewed another way, $500 million annually in foreign exchange cash for the colony.

Fun is always a strong motivator, but a more compelling reason that might send some tourists to Mars could be to save their lives. Currently medical research on Earth is hamstrung by various medical authorities. For example, during the recent COVID-19 epidemic, pharmaceutical companies came up with effective vaccines within weeks of the pandemic alert (Moderna took *two days* to develop its shot) but permissions to use them were not authorized by the Food and Drug Administration and comparable European authorities for almost a year. In the case of less public emergencies, permissions to employ various drugs and medical procedures have been delayed many years (as happened with the AIDS epidemic in America in the 1980s, when the US government denied access to lifesaving drugs in general use in Europe). Individuals facing death because of such bureaucratic stalling might be willing to travel a very long way to reach a treatment center out of the control of such authorities. If a Mars colony were to develop such a cure, or simply offer facilities for the employment of unapproved treatments developed elsewhere, it could get a lot of visitors. This would not only benefit the colony and those wealthy individuals capable of paying the high price involved. It would also be a godsend to all the people of Earth, whose opportunities to avail themselves of new treatments would be accelerated by their demonstration on Mars and its ensuing undermining of the regulatory regime.

Once development of the asteroid belt begins, Mars will stand to benefit greatly from another class of tourists: asteroid miners on shore leave. If you are running a hospitality business, there are no better customers than people who have a lot of money and nowhere else to spend it. Many of the cities of the American West owe their existence to the bonanza they reaped from this phenomenon. There is no reason Martian cities should not do the same.

Luxury Goods

During his 1931 trial, the notorious bank robber Willie Sutton was asked by the judge why he robbed banks. "Because that's where the money is," Sutton replied.

Sutton had a point. Indeed, the same logic was used with extraordinary success by Elon Musk when he chose to target the first product of his Tesla electric car company at the super rich.

That's where the money is.

He's not the only one who has made a bundle off those willing to throw away cash for prestige artifacts. There are many brands of items ranging from wristwatches to handbags that sell for an enormous markup over completely equivalent items, simply because the purchasers enjoy the prestige that comes from owning something too expensive for the masses to buy. Diamonds can be made artificially and look as beautiful as any that come out the ground. But artificial diamonds don't count, because they are not as expensive as those produced in slave-labor mines in Africa.

So, why not produce industrial diamonds on Mars, perhaps tinting them in a peculiar way with some trace minerals, and export them to

Earth for sale at sky-high prices? Or include them as decorations on designer wristwatches made on Mars?

In 2020, people worldwide forked over $28 billion for diamonds whose price was sustained only by the prestige of their costliness. Why shouldn't Mars get a piece of the action?

Spectator Sports

In 2022, the global market for spectator sports was $164 billion, and it is expected to reach $600 billion by 2031. Sports on Mars will be extraordinary. Because of the one-third gravity, basketball players will be able to jump three times as high, and there will probably be many new and very exciting sports invented that exploit the lower Mars gravity to even greater effect. For example, if arenas on Mars are pressurized to match that of Earth at sea level, players will be able to strap wings on their arms and fly like birds. This could make it possible to play sports comparable to the exhilarating Quidditch games portrayed in J. K. Rowling's Harry Potter novels.

Other spectacular sports could include cross-country rover races comparable to Alaska's Iditarod Trail Sled Dog Race.

Such sporting events could be transmitted to Earth and televised at considerable profit to mass audiences worldwide. If even a small fraction of the terrestrial spectator sport audience were to tune in, the income accruing to Mars could be considerable.

Material Exports

A developed Martian colony could also earn cash through material exports. In the short term, these exports will need to go to Earth, since that will be the only preexisting off-planet market. Later, Mars will be able to earn more by exporting to other destinations, including the Earth's Moon, the asteroids, and the outer solar system.

Let's start with exports to Earth.

A Starship will have a dry mass of a hundred metric tons and a propellant capacity of a thousand tons. Its Raptor engines each have an I_{sp} of 370 s, implying an exhaust velocity of 3,626 m/s. The velocity change, of ΔV, needed to take off from Mars and fly directly back to Earth is 6,500 m/s.

If we plug these numbers into the rocket equation given in Chapter 3,[2] we can see that the mass ratio will be exp(6500/3626) = 6. So, if the hundred-ton Starship is carrying a hundred tons of cargo, it will need a thousand tons of propellant. (Because wet mass is the sum of the dry mass plus propellant, and 1200/200 = 6.) This, in fact, is the capacity it has.

It takes around 10 kWh to produce and liquefy a kilogram of methane/oxygen propellant on Mars. That being the case, it will take 10 million kWh of power to make the 1,000 tons (= 1 million kilograms) of propellant needed to fuel a Starship. If the cost of power on Mars is $0.20/kWh (twice the current US average), then the power needed to refuel a Starship on the Martian surface will cost $2 million. If the Starship costs $20 million, at 5 percent interest it will cost $1 million per year to own a Starship. That adds up to $3 million for capital and fuel costs to send a cargo Starship back from Mars to Earth. To that we need to add pay for the crew and ground support team, insurance costs, repair and maintenance costs, renting hangar space at the spaceport, and so on, and also provide some margin of profit for the spaceline operator. Allowing for all this, let's take $10 million as the cost for the entire operation to send a cargo Starship from Mars to Earth. (The cost of getting the Starship to Mars was covered by the outbound passengers and freight. If the emigrants picked up a used beater from Astrolinea Mexicana for $5 million and sold it on arrival at Mars for $20 million, they could turn a small profit on their trip.) Divided by a hundred tons, that works out to a freight charge of $10,000 per ton, or $100/kg. Are there items that could be sent from Mars to Earth worth more than that?

You bet there are.

Let's start with deuterium.

As noted earlier, deuterium is about five times more common in Martian water than it is on Earth (833 ppm on Mars versus 166 ppm on Earth). A Mars city-state of 100,000 people will need to electrolyze at least a hundred metric tons of water per day, within which there will be 11.1 tons of hydrogen, from which 18.5 kilograms of deuterium can be extracted. The electrolysis power to do this will require about 480 MWe/hrs, or 20 MWe all day long. The amount of deuterium, however, will be enough to produce 30 GWe, if we assume the fusion energy is converted into electricity at 40 percent efficiency.

Thirty gigawatts (30 GWe) is enough to provide all the residential electricity currently needed by a US state with a population of thirty million. It's vastly more than what a Martian colony of one hundred thousand would need. So, what should be done with the surplus? That's simple—export it.

Deuterium currently sells for about $4,000/kg, which is forty times the $100/kg threshold needed to justify shipping cargo from Mars to Earth. If we wanted to send one ship's worth every 2.5-year synodic cycle, we would need to produce the deuterium at a rate of 110 kg/day, about six times that produced naturally by the life support system of our 100,000-person city. No matter, since there is money at stake, the town will be happy to expand its production for the export market. Each shipment will bring in $400 million in revenue and provide sufficient energy to meet the electricity needs of 180 million people on Earth.

There are plenty of other materials that are worth more than $100/kg. Here is a list of a few.

TABLE 5.1 Price of Selected Metals (November 7, 2021)

Metal	Price ($/kg)
Neodymium	$164
Gallium	$354
Dysprosium	$373
Silver	$857
Germanium	$1,500
Samarium	$3,339
Scandium	$5,735
Ruthenium	$21,875
Platinum	$36,750
Gold	$63,700
Palladium	$70,000
Iridium	$157,500
Rhodium	$525,000

No major finds of any of these materials have been made on Mars. But there is every reason to believe that they exist. In fact, the geology of Mars has been compared to that of Africa, which is particularly rich in such resources.[3] Moreover, unlike Earth, which has been vigorously searched for the past five thousand years for gold and silver, Mars is virgin territory for miners and prospectors. The obvious outcrops of such precious metals are still there. If they can be found, their value would more than justify the shipping cost.

Developing Phobos

The ability of Mars to export for profit can be greatly enhanced by the development of its inner moon, Phobos. Using Starships, colonists will be able to lift such resources from the surface to Phobos, where an electromagnetic catapult could be emplaced capable of firing the cargo off for interplanetary export. Larger or more complex

FIGURE 5.1. *A Phobos station could greatly improve the economics of Martian exports.*[4] *(Credit: Michel Lamontagne)*

cargoes could be shipped out from Phobos at low cost using robotic solar sail-powered spacecraft. The ΔV to reach Phobos from the Martian surface is not much different than that needed for flight to Earth. But flying from the surface to Phobos can be done every week, not every two years, allowing more use to be gained from every Starship involved in the trade. This will greatly reduce Starship capital costs as a component of export shipping expenses.

A much more dramatic improvement in the economics of Martian exports could be achieved by hanging a "skyhook" tether 5,800 kilometers down from Phobos. The little moon orbits at a speed of 2.138 km/s 6,000 km above the Martian surface, which itself is 3,400 km from the planet's center. This being the case, the bottom of the tether would be traveling at a speed of just 0.82 km/s above the Martian equator, which itself is turning in the same direction at a speed of 0.24 km/s. As a result, a rocket vehicle taking off from the equator could reach the tether with a ΔV of just 0.58 km/s, catch it, and then be hauled up to Phobos by a cable car. If there were another cable extending outward from Phobos, the vehicle could be *lowered outward* (sic— along this cable, effective gravity would point away from Mars, since centrifugal force would be greater than gravity) to reach escape velocity at 3,726 kilometers beyond Phobos, and still greater speeds at distances beyond. Assuming a tether tensile strength equal to Kevlar

(2,800 MPa), the tether would have a mass less than ten times the payload it could lift. By using such a system, cargoes could be sent back to Earth or whipped out to the asteroid belt or beyond with a tenth of the rocket ΔV required to do so directly. This would cut export propellant costs by an order of magnitude and make it feasible for Mars colonists to transport goods cheaply not only to Earth and its Moon, but to the asteroid belt, and the moons of Jupiter, Saturn, Uranus, and Neptune.

The Space Triangle Trade

The main asteroid belt is known to be filled with small bodies that are extremely rich in platinum group metals. But large-scale human activity in the main belt to exploit these potential resources will be difficult to support until there are human settlements on Mars. This is so because, while water and carbonaceous material can readily be found among the asteroids (making them as a group far richer than the Moon), it is not necessarily the case that such volatiles can be found on those asteroids that are most rich in exportable metals. Quite the contrary, the metal-rich type M asteroids are nearly volatile free. Moreover, while many of the main belt asteroids contain all the carbon, hydrogen, and oxygen needed to support agriculture, nitrogen is generally rare. Moreover, sunlight in the main belt is too dim to support agriculture, which means that plants would have to be grown by artificially generated light. This is a significant disadvantage for asteroid colonization, because plants are enormous consumers of light energy, and it is doubtful that growing plants with electric lights to support any significant population is practical with current space power sources. Moreover, while collectively the asteroids may someday possess a significant mining workforce, until very advanced and inexpensive robotic technology becomes available, it is unlikely that any one asteroid will have the sufficient manpower required to develop the division of labor necessary for true multifaceted industrial development.

Mining bases in the asteroid belt are a relatively near-term proposition. But farms, industries, and cities will need to wait until the widespread use of controlled fusion makes very large-scale employment of artificial power possible in the main belt. For the twenty-first century, most of the supplies needed to support the asteroid prospectors and miners will have to come from somewhere else.

As I first showed in detail in my book *The Case for Mars*,[5] even before a Phobos tether system is created, the ΔVs required to reach the asteroid main belt from Earth are more than double those required to access it from the Red Planet, leading to mass ratios many times greater and mission gross liftoff masses *fifty times larger*—and this is true whether chemical or electric propulsion systems are used. It may be noted that there is a class of asteroids known as near Earth objects (NEOs) orbiting between Earth and Mars. But the main belt is vastly richer, sporting hundreds of times more asteroids larger than one kilometer in size. Moreover, while it is considerably easier for Earth-based craft to reach these objects than the main belt, reaching them from Mars is even easier still. And once the Phobos tether transportation system is put in place, the Martian payload delivery advantage increases an order of magnitude. I've now reworked the numbers governing asteroid logistics based on the selection of Starships as the interplanetary transportation of choice. The results of this analysis are presented in the focus section following this chapter. They are even more forceful.

The conclusion that follows is simply this: Anything that needs to be sent to the asteroid belt that can be produced on Mars will be produced on Mars. High technology goods needed to support asteroid mining may have to come from Earth for some time. But since food, clothing, and other necessities can be produced on Mars with much greater ease than would be possible anywhere farther out, Mars will become the central base and port of call for exploration and commerce heading out to the asteroid belt, the outer solar system, and beyond.

The outline of mid-future interplanetary commerce in the inner solar system thus becomes clear. There will be a "triangle trade," with Earth supplying high-technology manufactured goods to Mars; Mars supplying low-technology manufactured goods and food staples to the asteroid belt and possibly the Moon; and the asteroids sending metals and perhaps the Moon sending helium-3 to Earth. This triangle trade is directly analogous to the triangle trade of Britain, her North American colonies, and the West Indies during the colonial period. Britain would send manufactured goods to North America; the American colonies would send food staples and needed craft products to the West Indies; and the West Indies would send cash crops such as sugar to Britain. A similar triangle trade involving Britain, Australia, and the Spice Islands also supported British trade in the East Indies during the nineteenth century.

Miners can make a lot of money off precious metals. But the people who really profit the most from any gold rush are those who sell the miners what they need.

The Martians will do well.

FOCUS SECTION

WHY MARS WILL GET THE
ASTEROID TRADE MONOPOLY

I've said that Mars enjoys an enormous positional advantage over Earth for supporting lucrative mining operations in the main asteroid belt. In this section, I'll show you the math that makes it so. (If you don't like math, that's okay. Just take my word for it and skip to the next chapter.)

The amount of propellant a rocket needs to go somewhere is defined not by the distance to that destination, but by the velocity change, or ΔV, needed to get on a trajectory to that place. In Table 5.2, I show what these ΔVs are for travel from Earth or Mars to the asteroids, as well as the mass ratios they imply. In all cases except for the final entry, I assume the ship is using Starship-type propulsion with a specific impulse of 370 s. The option of using advanced nuclear electric propulsion (NEP) with a specific impulse of 5,000 s is considered in the last entry.

TABLE 5.2 Transportation in the Inner Solar System

	Earth		Mars		Mars with Phobos Tether	
Maneuver	ΔV	Mass Ratio	ΔV	Mass Ratio	ΔV	Mass Ratio
Surface to low orbit	9.3 km/s	13.0	4.0 km/s	3.0	0.7 km/s	1.21

	Earth		Mars		Mars with Phobos Tether	
Maneuver	ΔV	Mass Ratio	ΔV	Mass Ratio	ΔV	Mass Ratio
Surface to Escape	12.4	30.7	5.6	4.70	0.7	1.21
Low Orbit to NEO	4.0	3.0	1.6	1.56	0.7	1.21
Low Orbit to Ceres	**9.6**	**14.2**	**4.9**	**3.87**	**2.9**	**2.22**
Near Escape to Ceres	**6.7**	**6.37**	**3.5**	**2.63**	**2.9**	**2.22**
Surface to Ceres	18.6	170.4	8.9	11.7	3.6	2.70
Ceres to Planet	4.8	3.77	2.7	2.11	2.7	2.11
NEP:RT orbit to Ceres	40	2.66	15	1.36	13	1.30

Before we dive into the data in Table 5.2, we should note that the Starship has a dry mass of a hundred tons and a propellant capacity of a thousand tons. That means the maximum mass ratio it can muster, with zero cargo, is 11 (because (100 + 1000)/100 =11). If it is carrying a hundred tons of cargo, the best it can do is a mass ratio of 6.

Now look at Table 5.2. To go from the Earth's surface to orbit requires a mass ratio of 13. That is more than the Starship can do by itself, which is why it needs to be lifted by a Super Heavy first stage to make orbit. That's costly, but doable. However, now look at the mass ratio involved in going from low Earth orbit to Ceres, in the heart of the main belt. It's 14.2! So, even with zero cargo, the Starship can't manage that. In contrast, going from Mars orbit to Ceres, even with no Phobos tether, the mass ratio is just 3.87. A Starship with its tanks full could do that carrying 248 tons of cargo! As if that were not enough, once the Phobos tether is in place, the Starship could do the maneuver carrying 720 tons of cargo!

The cargo capacity associated with all the ΔVs listed in the table can be seen in Figure 3.3.[6]

For the Starship to go to Ceres from Earth, it would first need to be boosted from low Earth orbit to near-escape-velocity, high-energy condition, like that of a trans-lunar injection orbit, a LaGrange point, or NASA's proposed Deep Space Gateway. From there it could go to Ceres with a mass ratio of 6.37, meaning it could do the trip carrying eighty-six tons of cargo. But getting it to the Gateway from LEO requires a mass ratio of 2.3, which means that it would need to be refueled with 240 tons of propellant in Earth orbit, which would require orbital refueling by three tanker Starships. But that would just get it to the Gateway with zero propellant. Once at the Gateway, it would need to be refueled by ten more tankers, which would each need refueling by three more tankers launched from Earth in order to reach the asteroid-exploration Starship. That's a total of forty-four Starship launches just to support sending one carrying a mere eighty-six tons of outbound cargo from Earth. But wait, there is more: The ΔV for the ship to return to Earth is 4.8 km/s, which requires a mass ratio of 3.77 to execute. That means it would need 277 tons of propellant just to get back to Earth, even if it were carrying zero cargo (which would be an unsatisfactory outcome for a mining expedition). So, unless it could be refueled by suppliers in the belt, its mission would not merely be ruinously expensive, it would be impossible.

In short, Starship-type vehicles based on Earth can't support asteroid mining in the main belt, while those on Mars can do so quite readily. But what if technology advances? Nuclear electric propulsion is a future possibility. The prospect of asteroid miners being able to afford multi-megawatt nuclear electric spaceships is questionable, but let's put that issue aside. Using such a ship—or a fusion-propelled one—specific impulses of 5,000 s or more are obtainable. You wouldn't want more for this mission, however, because the thrust of a nuclear engine of a given power goes down in inverse proportion to its exhaust velocity. As shown in Table 5.2, the ΔVs associated with the low-thrust trajectories used by electric or plasma propulsion systems are significantly greater than those that high-thrust, chemical rockets face.[7] Nevertheless, the very high exhaust velocities achievable by advanced propulsion systems make the resulting mass ratios manageable. However, the mass ratio of 2.66 of the nuclear electric propulsion spacecraft leaving Earth still implies that such a ship would need to use five times the propellant of one leaving Mars with a mass ratio of 1.36 (because the propellant fraction of a rocket is equal to the mass ratio minus 1, and 2.66 - 1 = 1.66, which is five times 1.36 - 1 =0.36). This means that a spaceship employing an engine of the same size to voyage to the asteroid belt from Earth will take five times longer than a Martian ship to do the trip and expose its engine to five times the wear and tear to make the same delivery.

So, there is no way around it: Mars will get the asteroid miners' business.

6

TRANSPORTATION
ON MARS

HAVE SHOES, WILL TRAVEL. The best way to get around town on Mars will be on foot.

Martian cities will be designed as relatively compact towns—ideally with an interesting street scene that makes pedestrian travel congenial—and promote opportunities for exercise to help their citizens keep in shape in Mars's one-third gravity environment. Those wishing to travel faster than walking could simply run or make excellent use of roller skates. For longer-distance travel within the city, bicycles will be more than sufficient, since with Mars's lower gravity cyclists could obtain triple the speed they can on Earth with equivalent effort. This will also make possible human-powered flight, since even in a 5-psi (pound-force per square inch) atmosphere, a Martian cyclist traveling at sixty miles per hour (if you can do twenty miles per hour on your bike on Earth, you could do sixty miles per hour on Mars) would only need to spread about two square meters of wing to take off and glide. For the same reason, human-powered tricycle delivery vehicles will also be quite practical for intra-city shipping, with only the largest cargoes requiring electric trucks.

For travel in nearby areas outside the city, electric vehicles of various sorts, including rovers utilizing wheels, half-tracks, treads, or even legged mobility systems, will do fine. But Martians will also need technologies for true long-distance travel. This need will become apparent well before the first cities are built, as the most powerful technology for improving the cost-effectiveness of a Mars exploration program is enhanced surface mobility. A Mars exploration expedition with a range of five hundred kilometers could explore a hundred times the area as one limited to a fifty-kilometer reach.

Mars terrain can be very rough, with transverses sometimes blocked by boulder fields or impassible canyons. For this reason, flight systems will be developed to aid Mars exploration. This has already begun with the demonstration of the subscale *Ingenuity* helicopter on the 2021 *Perseverance* Mars mission.

Ingenuity was very small, with negligible payload capacity. But just as the successful demonstration of the toy-car-sized *Sojourner* rover on the 1997 *Pathfinder* mission led to the go-cart-sized *Spirit* and *Opportunity* rovers in 2004 and then to the Volkswagen-sized *Curiosity* and *Perseverance* rovers in 2012 and 2021, so we can expect much larger and more capable exploration helicopters to take off following *Ingenuity*'s example. This could eventually lead not only to terrific airborne exploration rovers offering hundreds of times the science return of slow-moving ground systems, but even transports for human explorers.

While offering several advantages for exploration, helicopters are not attractive for cargo transport. This is because helicopters must generate all their lift directly by brute power. In contrast, a winged aircraft generates its lift using aerodynamic forces, and only needs to use its own power to keep moving forward by providing enough thrust to match its drag. An airplane's required thrust is thus lower than its lift by a factor known as its "lift over drag ratio" (L/D). On a typical subsonic airplane, like an airliner or a Cessna, the L/D can often be about 10. This can be increased to as much 20 on a sailplane, while supersonic jet fighters may be able to get around 4.

It is true that even the most efficient air cargo systems are vastly more expensive per kilogram delivered than trucks, which themselves are beaten silly by trains when compared on an energy cost-per-ton-mile basis.

Trucks require roads, or at least trails, which must be cleared. Trains require rails, which take not only work but a lot of finished materials to build. So, truck convoys moving along dirt roads will probably come early to enable delivery of ores or other materials from the countryside to the city. But the advantages of railroads are so profound that such systems will eventually be built to allow large-scale trade between Martian cities separated by continental distances. This will offer great benefit to settlements everywhere, since no one location on the planet

FIGURE 6.1. *May 10, 1869—The transcontinental railroad is completed, enabling coast-to-coast travel in days instead of months, and opening huge coastal markets for the settlers of America's interior. A similar great day awaits the pioneers of Mars. (Credit: Wikipedia Commons)*

is simultaneously rich in all the resources of interest. As a preliminary step toward implementation to minimize the required industrial capacity needed for railroad construction, cable-car-type systems taking advantage of Mars's low gravity and the advent of very high tensile strength materials like Kevlar and Spectra could be strung out over long distances. If the pylons contained motors to raise or lower the cables, cargoes could be sent scooting along the lines much as zip-liners do on Earth. But the advent of true steel railroads on Mars—perhaps using nuclear-powered locomotives—will be as transformative for the development of the Martian frontier as the completion of the transcontinental railroad was to the opening of the American West.

Flying on Mars

But until the trans-Mars railroads are built, intercity transport on the Red Planet is going to need to fly. There are several ways this could be done.

The Mars atmosphere at "the datum," the equivalent of Mars's sea level, has a density of 16 gm/m^3, about the same as Earth's at an altitude of thirty-one kilometers (one hundred thousand feet). However, when you consider Mars's lower gravity, winged flight on Mars presents the same difficulty as Earth does at twenty-four kilometers (seventy eight thousand feet). There are two airplanes that can fly that high: the high L/D subsonic U-2, and the supersonic SR-71.

As noted, subsonic aircraft can have higher L/D ratios than supersonic, with slow-moving albatross-like gliders (and albatross-like birds) with very long wingspans having ratios as high as 20. The total energy needed to achieve a flight of a given distance goes down in inverse proportion to the L/D. For this reason, several engineers, most notably crack airplane designer Dan Raymer,[1] have chosen a U-2-style high L/D subsonic aircraft as the basis for long-distance travel on Mars.

FIGURE 6.2. *Two ways to fly in Mars-thin air. Upper: the subsonic U-2 (Credit: US Air Force). Lower: the supersonic SR-71 (Credit: US Air Force and NASA).*

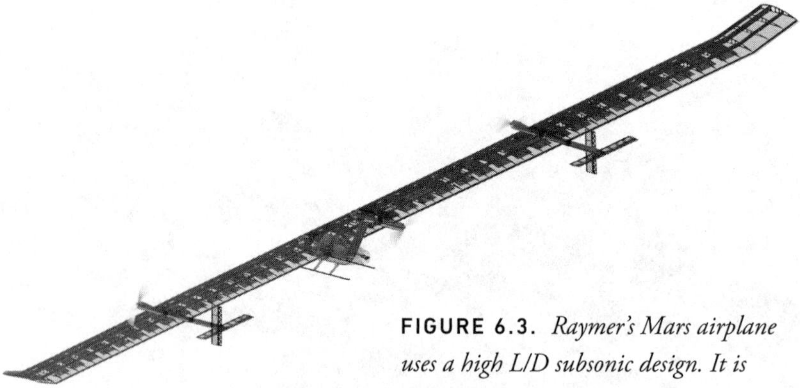

FIGURE 6.3. *Raymer's Mars airplane uses a high L/D subsonic design. It is designed for a crew of two Mars explorers, but much larger versions could be adapted for cargo. (Credit: Dan Raymer)*

Raymer's airplane uses vertical rockets for takeoff, due to Mars's "deplorable lack of paved runways." Martian cities would correct the latter deficiency, making conventional runway takeoff possible.

The speed of sound on Mars is about 240 m/s (540 mph). So, the fastest a high L/D subsonic plane could fly on Mars would be about 200 m/s (430 mph), or Mach 0.8. Flying at this speed a few km above the datum, a Mars airplane with a mass, including cargo, of 100 metric tons would need a wing area of 2,000 m². To get a high L/D, its wingspan would have to be more than ten times its wing chord length, so perhaps 150 meters by 14 meters.

That's a lot of wing. But the amount of wing area an airplane needs decreases in inverse proportion to the square of its air speed. So, if instead of traveling at Mach 0.8 the plane traveled at a supersonic Mach 4 (as does the SR-71), the required wing area would be cut by a factor of 25, and it would only need eighty meters² of wing. Since the plane is supersonic, this would come in the form of a delta wing, about eighteen meters long and spanning around nine meters at its base. That could be a lot more manageable. For comparison, the Starship is fifty meters long. So, one way such a vehicle might be designed could be by attaching delta wings to Starships after they arrive on Mars, thereby converting them into supersonic rocket planes.

Starships could also be used without wings as ballistic hopper vehicles for point-to-point transportation. But for short- and medium-distance travel (up to several thousand kilometers), the range achieved for a given amount of rocket propellant by winged systems is better.

The math on this is complicated. Those interested can find much of it in note 2. But the results are easy enough to understand. I show these in Figure 6.4 below.

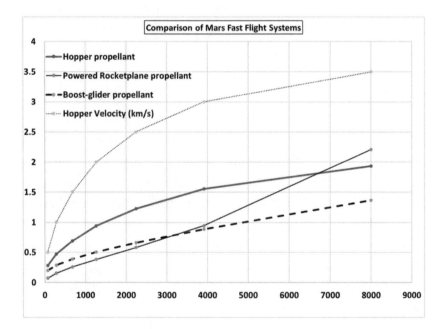

FIGURE 6.4. *Comparison of performance of fast flight systems on Mars. The horizontal axis is the flight range in kilometers. The vertical axis shows the amount of propellant needed as a fraction of the flight vehicle's dry mass for a ballistic hopper, a boost-glide system, and a powered rocket plane, both with an L/D of 4. Methane/oxygen propulsion with an I_{sp} of 370 s is assumed. Also shown is the hopper flight vehicle's velocity. The powered rocket plane's speed is half of this, while the boost-glide system speed is half of the hopper's at short range, rising to 77 percent at the eight-thousand-kilometer range.*

In Figure 6.4, we consider both powered rocket planes that, after runway takeoff to flight velocity, keep their engines running at low power to maintain flight speed, and boost-glider systems that launch vertically to high speed and then glide for the rest of the flight.

Looking at Figure 6.4, you can see that if the propellant fraction is 0.5 (i.e., fifty tons of propellant for a hundred-ton vehicle), the powered rocket plane can fly 2,000 kilometers, much more than either of the other concepts. However, if the propellant fraction rises to 1.0 or greater, the boost-glider system is superior, able to go 5,000 kilometers (roughly the distance from the Mars equator to its pole) using tons of propellant, compared to 4,000 km for the powered rocket plane, or 1,500 km for the hopper.

In short, for long-distance, surface-to-surface travel, winged supersonic rocket planes are the way to go. If the mission is to reach orbit, however, the wings are just extra weight. So, both ballistic hoppers and rocket planes will have their place on Mars, with each best for its chosen mission but offering some capability for that of the other.

Using Starships—either as is or modified with wings—for long-distance travel on Mars is attractive because large numbers of them will be available after their use for one-way transport from Earth by immigrants. But one problem with this concept is that such vehicles will need to refuel at the destination to do round trips. This is not a critical issue for intercity travel, because there will be refueling facilities at the destination. For round trips from the city to the outback, a system that can refuel itself would be desirable.

Fortunately, there is a potential propellant available everywhere on Mars that is much easier to produce than methane/oxygen. Mars's atmosphere is composed of 95 percent carbon dioxide, and if compressed to 10 bar (150 psi) at Martian ambient temperatures, it will liquefy. This is the first step in the process of making methane/oxygen rocket propellant on Mars out of carbon dioxide and water, but it is much simpler and only requires less than a tenth of the energy needed to do the rest. A Mars flight vehicle could readily carry the gear (a roughing pump) and the power supply needed to tank itself up with liquid carbon dioxide drawn right out of the air.

Now, pressurized, raw carbon dioxide would be a poor propellant if used cold. It would have the exhaust velocity of a fire extinguisher (which is basically a raw carbon dioxide rocket), yielding an I_{sp} of perhaps 60 s. But if the carbon dioxide is first passed through a nuclear thermal reactor and heated to 2,500° C before being exhausted out of a rocket nozzle, a specific impulse of 264 s can be obtained—a bit better than that of the V-2 rockets used by the Germans in World War II. While significantly inferior to the 370 s performance you can get using methane/oxygen, it's good enough to fly.

I developed this concept, which I called the "Nuclear rocket using Indigenous Martian Fuel," or NIMF, while I was working at Martin Marietta Astronautics in the late 1980s.[3, 4] Nuclear thermal rockets (NTRs) were developed and successfully tested by NASA under the Rover and NERVA programs in the 1960s using hydrogen propellant.[5] This was done because hydrogen is much lighter than the water vapor exhaust produced by a hydrogen/oxygen rocket. By heating it to comparable temperatures in a nuclear reactor, an exhaust velocity twice that of the best possible chemical rockets can be obtained. But the same general approach—using a nuclear reactor to apply heat to an inert gas to create a flying steam-kettle—can be used with other fluids. Instead of looking for the highest performing NTR propellant, I decided to look for the cheapest, and on Mars, carbon dioxide is as cheap as air. Thus, the NIMF.

NIMF vehicles can be designed in the form of winged rocket planes or ballistic hoppers. They will take some work to create, because the carbide fuel elements used in NERVA engines would react with carbon dioxide at high temperatures. Instead, uranium-thorium oxide fuel elements would need to be developed, and probably used in a particle or pellet bed reactor design to maximize their surface area, and thus potential thrust. The effort would be worth it, however, because NIMF technology would endow Martian settlers with long-range transports that can refuel themselves each time they land. Furthermore, the oxide nuclear fuel pellets created for use with carbon dioxide propellant on Mars could work well with water propellant. NIMF vehicles could thus also serve as workhorses for asteroid miners working in the main belt, as water ice is abundant there.

FIGURE 6.5. *NIMF Vehicles. Left: NIMF rocket plane using Harrier-type ventral vertical takeoff system before starting its main engine to begin horizontal flight. Right: NIMF ballistic hopper. (Credit: Robert Murray, Lockheed Martin)*

The Mars Tether Authority

Once the Phobos tether system is in place, Martian rocket planes or hoppers could use it to great advantage to travel around the planet with greatly reduced propellant consumption. As noted in Chapter 5, the base of the Phobos tether would be moving just outside the atmosphere in an equatorial orbit at a velocity, with respect to the ground, of just 0.58 km/s. A rocket vehicle located on Mars's equator could take off and perform a suborbital rendezvous and latch onto it with a ΔV not much more than this. It could then be hauled up the tether by cable car and reach stable orbit at Phobos or go beyond on the outbound cable and be released into a high energy orbit that could take it to Earth or the asteroid belt, as desired. But there is another alternative: It could just hold on to the bottom of the tether and get a free ride around Mars's equator and then let go anywhere along its path it cared to land. For total ΔV of less than 1 km/s (including 200 m/s for launch and rendezvous hover and 200 m/s to land), which ordinarily would allow a hopper to fly a mere hundred kilometers, it could travel halfway around the planet! That's a ground track distance of ten thousand kilometers, *a hundred times* what it could do on its own using the same amount of propellant.

That's great, but as the astute reader may have already observed, such a transit system would have a key limitation: It could only serve customers on the Martian equator. Of course, if you live close to the equator, you could drive there to catch the Mars Tether Authority line, which would fly over your local stop every 11.2 hours, or 2.2 times per sol. (It actually orbits the planet 3.2 times per sol, but since Mars is turning once every sol, it would only seem like 2.2 times per sol to you.)

But the MTA should be able to offer better service than that, and there is a way it could. Phobos may be a nice place to visit, but frankly, *it's in the wrong orbit*. Why should the MTA limit itself to operations based on a poorly designed natural moon? It should not. Instead, it should build a properly designed space station of its own.

The best place to put the MTA station is in a near polar orbit. That way it would be able to provide travel service to the entire planet. If we put it in a Phobos altitude circular orbit inclined at a 73-degree angle to the equator, its west-to-east ground track velocity would precisely match the planet's rotation, and a rocket taking off from anywhere on the planet between 73 degrees N and 73 degrees S would be able to catch its line with a north–south ΔV of 0.78 km/s. Like Phobos, it would orbit the planet 3.2 times per day, but appear to groundlings to be orbiting 2.2 times per day. Traveling first south and then north, it would cross every latitude line between 73 degrees N and 73 degrees S 4.4 times per day. If we wanted to regularize service to exactly four crossings every day, we could do so by putting it in a 6,400-kilometer altitude orbit, just slightly higher than Phobos's 6,000 kilometers. But this would limit it to distinct ground tracks separated by 90 degrees of longitude getting regular service twice per sol, with those offset from these paths getting none. By *not* synchronizing its orbit with Mars rotation, it would be able to offer service, albeit less frequently, to everyone, everywhere. To increase the frequency of service, the MTA could build multiple stations, with their orbits set at phased angles to one another. Like the Phobos station, each station should have both a down tether enabling access to orbit, and an up tether providing boost to interplanetary trajectories.

Then Martians on a budget will be able to travel anywhere to anywhere on Mars, or anywhere on Mars to anywhere in the solar system, riding the MTA.

7

CITIES ON MARS

Well building hath three conditions: firmness, commodity, and delight.

—Vitruvius, Roman Architect, circa 20 BC

MARTIAN CIVILIZATION will be centered on cities. There are many reasons why this will be so.

Basic physics, for one, argues for housing people on Mars in cities rather than large numbers of individual habitations. Livable space on Mars will have to be created in the form of pressurized and temperature-controlled volumes walled off from the outside Mars environment. The bigger anything is, the greater will be the ratio of its volume to its surface area. To create a given volume of pressurized and temperature-controlled space, it takes a lot less wall if the volume is enclosed in a single barrier, such as a city dome or underground cavern, rather than thousands of individual habitations scattered across the planitia (low plains). The smaller the surface area, the less heat will be leaked from the habitable space to the cold world outside. This means that less total power will be needed. More important, the cost of a power system, such as a nuclear reactor, typically goes up, not in proportion to its size, but as the square root of its output. It might take twenty kilowatts to power a home for an individual family living alone out on the planitia, but only ten kilowatts per family living with others in a single city. Living in scattered homes, a thousand families would need a thousand twenty-kilowatt reactors, each costing, say, one million Martian dollars each, for a total of a billion Martian dollars. Ten thousand families living together in the city would require a single hundred-megawatt reactor for their residential power. This would cost $70 million dollars in total (because $1 million $(100,000/20)^{0.5}$ = $70 million), or $7,000 per family, which would be much more affordable. The price-to-capacity ratio of other necessary systems, such as life support and waste disposal, scale in similar fashion. In addition, the cost of home construction, communications, material deliveries, and all other utilities per family will be much less within the city compared to outside. Economies of that magnitude will be quite compelling.

But there is another reason Martian civilization will be composed of cities. Mars's most important export will be inventions, and technological innovation cannot be done alone. I am a fairly prolific inventor myself, but I would not have been able to get very far without the assistance of my team, which includes mechanical and electrical engineers, machinists, chemists, support from financiers, sales types, and lawyers, and perhaps most critically, access to vendors. No matter how broadly and deeply skilled you may be, to invent anything, you need to be able to get your hands on all sorts of parts, materials, and tools made by lots of other people. Furthermore, almost all inventions are combinations of previous inventions and ideas, and these all come from other people. Interaction with others is fundamental to the whole inventive process, starting with perception of the problem that needs to be solved, gaining ideas that will lead to the problem's solution, making the invention, testing and proving it under conditions that others will demand, getting others to see its merit, and bringing it to market. In short, invention is a social process. It is best done in cities.

Industrial production of all complex items is also a social process. That is why the Industrial Revolution led to the urbanization of terrestrial society. Mars cities will also be centers of certain kinds of industry, making the machines and materials needed for the cities themselves, as well as for export to out-of-town mining and agricultural outposts. As on Earth, most industries require a division of labor, including workers of every category, as well as vendors and their support personnel who can only be found in cities.

Finally, humans are social creatures. While we may enjoy brief moments of solitude, it is only in the company of others that we can experience life in full. Prolonged isolation is a form of punishment—and used as such in prisons. Preindustrial societies, while primarily rural in nature, nevertheless sought to avoid such hardship by clustering in villages. Then, once the Industrial Revolution began, affording many the opportunity to leave their villages for larger towns and cities, few chose to come back.

It is very unlikely that many people would want to leave Earth for a life of solitude out on the Martian planitia. If Mars is going to attract immigrants, it will need to offer urban environments with a dynamic social life.

How ya gonna keep 'em down on the farm, after they've seen Paree?

So, cities it will be.

On Earth, there are two types of cities: foot cities and car cities. The foot cities include the more charming European cities that survived World War II, as well as parts of those American cities such as Boston, New York, Philadelphia, San Francisco, and Seattle that were born and grew before the advent of the automobile. The car cities are places like Los Angeles that are largely creatures of the automotive age. In foot cities, people travel by foot, with neighbors frequently meeting one another on the sidewalk as they move about from house to store, café, and pub. In car cities, people travel by car, zipping about from home to one strip mall after another, rarely interacting. When people from foot cities go to a car city, they have a hard time finding it. To them, the car city seems not to be a city at all, but rather some kind of gigantic suburb, without the urb. As *New Yorker* wit Dorothy Parker once said of Los Angeles, "There's no there there."

Martian cities will be foot cities.

Old European cities were foot cities in part because they needed to fit inside fortress walls for protection against enemies. Mars cities will need to fit inside walls to protect themselves from the hostile Martian environment. As was the case with medieval-born European cities, Martian urban population density will be relatively high.

Suburban sprawl is out of the question for Mars for energetic reasons, as well. Inhabitants of car cities like Los Angeles use an enormous amount of energy driving long distances from one strip mall to another at high speed in multi-ton vehicles. Work is force times distance. To first order, someone driving a 1,500-kilogram car ten kilometers to shop for something expends a thousand times as much energy as someone with a mass of sixty kilograms walking 250 meters to pick up a package at the store down the street. On Earth, we can get away with this because both vast quantities of fuel and the oxygen with which to burn it have been left for our use by billions of years of photosynthesis. Martians will need to put a higher value on their power, because it will both be in more limited supply and much greater demand.

Circumstances may force Martians to live in foot cities, but that may ultimately prove to be a great advantage. The most important business of Mars cities will be invention, which is greatly facilitated by the kind of human social interaction that occurs naturally in a foot city. To be sure, urban design is not the only factor, or even the most important factor, in the inventiveness of a society. Culture matters much more. Many more inventions come out of Los Angeles than Lagos. But on a per capita basis, Boston is far more creative still.

Logistical Requirements for a Mars City

Martian cities need not be multimillion-person metropolises like modern New York or Paris. Such a giant city would be very hard to supply before extensive, cheap, long-distance rail transport is brought into being on Mars. Furthermore, it is unnecessary. Some of the most creative cities in human history, like Renaissance Florence or Amsterdam, or Philadelphia at the time of the American Revolution, had populations on the order of fifty thousand. Let's take this as our baseline for the size of a typical early Mars city.

SPACE

How big will the city need to be? That depends upon its population density. Table 7.1 presents the population density of several well-known cities on Earth.

TABLE 7.1 Population Density of Sample Cities

City	Population	Size (km²)	Density (persons/ha)
Manila	1,846,513	43	429
Paris	2,187,526	105	208
Barcelona	1,636,732	101	162
San Francisco	805,235	121	67
Boston	645,149	125	51
Amsterdam	851,573	166	51
Toulouse	479,638	118	41

Manila is the most densely populated city on Earth. It is not known as a particularly livable city, however, so 429 persons per hectare is probably too great. (A hectare is an area equal to a square one hundred meters on a side, encompassing about 2.5 acres.) At 200 person/ha, Paris is much nicer, while very pleasant cities like Boston and Amsterdam come in around 50/ha. On Mars, space will be at a premium, so, trading aesthetics against economics, perhaps 125 persons/ha (roughly midway between San Francisco and Barcelona) might be a reasonable design goal. This would imply a need for about 400 hectares (4 square kilometers, or 1.6 square miles) of pressurized living area for the town.

There are several alternative ways to create such living space. But before we get to construction, let's have a look at what will be required to meet the city's basic needs.

MATTER

The largest material necessities people require every day are water, food, and oxygen, in that order.

Water and oxygen will be recycled, but some water will be wasted. If we assume a wastage of ten kilograms of water per day per person, that amounts to a requirement to mine five hundred metric tons of new water every day. Small tanker trucks have a capacity of three thousand gallons, or eleven metric tons of water. This would be trivial to supply by pipeline, but even if the water needed to be trucked in from an ice mine some distance from the town, this would only amount to about forty-five truckloads per day. This would not be difficult for a fifty-thousand-person town to manage.

Food is a much more formidable requirement. Growing food requires space. The total amount of farmland on Earth is 1.6 billion hectares. Given the Earth's population of 8 billion people, this means that global agriculture currently supports the world's population at a rate of five people per hectare.[1] But world agriculture is extremely inefficient. Much of the land is used without irrigation or fertilizer, or as rangeland or pasture, or for growing nonfood or fuel crops like cotton or biofuels that can easily be replaced by chemical industry products.

A much better example to use as a basis of comparison would be Iowa corn farms, which in 2021 averaged about two hundred bushels of corn per acre, or twelve metric tons per hectare, per year. A kilogram of corn contains about 3,500 kcal (a.k.a. calories) of energy. This is about 1.5 times the recommended caloric requirement for an average person (2,500 for men, 2,000 for women). So, assuming no wastage, and no diversion of product for animal feed or other uses, one hectare of Iowa farmland can provide sufficient corn to meet the raw caloric needs of fifty people.

But why corn? It is possible to raise forty-five tons of potatoes per hectare per year, or almost four times the crop mass of corn. Unfortunately, potatoes only have an energy content of 930 kcal per kilogram, one-fourth that of corn. So, the dietary caloric yield per hectare is about the same. The same is true for wheat, sorghum, rice, and common beans. All of them should be grown, for the sake of variety, but the math dictating the required acreage does not change much.

Some of the food product will be wasted, however. Furthermore, only the poorest people on Earth subsist on a pure carbohydrate diet. People want and need fruits and vegetables as well, and preferably fish or meat, too. A variety of these will be necessary for a successful Mars city, not merely to meet minimum requirements for a healthy diet, but to attract immigrants used to having them on Earth. If the city wants to be able to draw immigrants from Asia, it will have to produce rice and Asian vegetables, but Europeans and Americans prefer wheat and dairy product, too. If the city wants both, it will need to grow for both. You can avoid scurvy by eating limes, but few would wish to move to a city where limes were the only fruit available.

Traditional agriculture requires not only land, but soil, which will have to be created on Mars by washing the perchlorates and other undesirable components out of the regolith and building up its organic content through a series of steps. Many fruits and vegetables can be grown without soil using advanced techniques such as aeroponics or hydroponics, which also produce a higher yield per hectare use than traditional agriculture. However, these techniques are both labor and technology intensive, which makes them more costly than traditional agriculture on Earth and may prove to be so on Mars as well. Their spatial efficiency makes them suitable for implementation in indoor vertical farms. But such systems require artificial illumination, which would put a heavy load on the city's power supply if done on a large scale.

Only a minority of people are vegetarians. A fair number of people are willing to get by with less meat, fish, or eggs than is customary in the current American diet, but few are willing to give them up entirely. Meeting this need, even with relatively efficient species such as tilapia fish and chickens that can make use of either aquaponic or land crop residues, involves loss of food chain efficiency.

People also greatly desire a variety of herbs and spices with their foods—so much so that at one time Europeans sent ships halfway around the world at great cost and risk to be able to procure them. Mars cities won't need to cater to such desires for a varied cuisine to simply *survive*, but those that do will rapidly outgrow the rest. So, from the point of view of natural selection, a less than maximally efficient varied agricultural sector will be a necessity.

It's true that certain foods can be produced by physical-chemical means that are potentially much more efficient than agriculture. This includes simple chemicals like sugar and ethanol. Using biotechnology, it should be possible to produce starch and grow cultured meats. Many writers have also discussed producing proteins by growing worms or insects on an industrial basis. Mass production of spirulina algae in vats is also possible and will probably be done as spirulina have a very high protein content.

But people are rather particular about their food. They don't just care about its nominal dietary value, but its taste and texture as well— as proven by the great difference in price people are willing to pay for different wines, or the markup of steak or salmon compared to bologna or fake fish.

Early pioneers can rough it and settle for whatever food can be produced most efficiently. But while industrial techniques may have their uses in reducing the total fraction of the diet that needs to be produced using traditional plants and animals, successful cities will probably need to produce at least half their food the old-fashioned way.

On the other hand, crop yields on Earth are limited by its seasons. No corn is grown in Iowa during winter. On Mars, crops will be grown in greenhouses or other indoor facilities unaffected by seasons. This will roughly double yields per hectare relative to Earth.

Taking all this into account, I believe an estimate of one hectare of farmland per forty citizens is probably in the right ballpark. This compares to that currently prevailing in Israel (twenty-five persons per hectare), South Korea, and Japan (thirty-three persons per hectare), all of which are effectively island nations in an agricultural sense, but which make use of some food imports.

At forty persons per hectare, our fifty-thousand-person Mars city will require 1,250 hectares, or five square miles of farmland. This brings us to the very important question of how it will be lit.

ENERGY

The human race currently uses about twenty terawatts (TW) of power, which, divided by eight billion people on the planet, works out to 2.5 kW per person. Most of humanity is rather poor, however. The United States uses 2.47 TW, or an average of 7.5 kW for each of its 330 million citizens.[2] This is divided between 37 percent for transportation, 35 percent for industrial, 16 percent for residential, and 12 percent for commercial uses. The transportation fraction is disproportionate, caused by the large size of the United States relative to its population, necessitating a great deal of energy-intensive transport using the interstate highway system and Americans' proclivity to travel everywhere by automobile. Martian foot cities drawing most of their resources from nearby will expend a much lower percentage of their power on transportation. On the other hand, Americans get their oxygen at zero energy cost and don't need to recycle their water. To meet these needs, Mars cities will require a life support system. This can be powered, however, with less than 0.5 kW per person (an average human burns oxygen at a day-night average rate of about 0.1 kW, while electrolysis units can produce oxygen with an efficiency as high as 85 percent). Given that, it could be argued that a Mars city could be powered at a per capita rate somewhat lower than the current US average, perhaps 5 kW/person. But energy consumption per capita has been increasing constantly for the past two hundred years, directly correlating with rising living standards as more types of machines have been invented. Fifty years ago, the amount of energy Americans used to power computers was insignificant. Today, about 6 percent of all energy used worldwide[3] is being used to drive computers and the internet, with a much higher percentage prevalent in the advanced sector. Martians may make extensive use of robots and employ many other energy-intensive technologies that have yet to make their debut. Taking all this into account, an estimate of 10 kW per person (for all types of power, not just residential) would seem to be reasonable. For a fifty-thousand-person city, that works out to a power requirement of 500 MW, roughly the output of a mid-sized nuclear reactor. This would probably be supplied by several smaller reactors, as the city would start small, and only add to its grid as it grew.

Small modular reactors currently under development could meet this need. For example, the Los Alamos–designed Hyperion Power Module liquid-metal-cooled reactor produces 25 MWe and has a mass of fifty tons, making it transportable to Mars in a Starship-type vehicle. Larger reactors could be assembled on Mars once the ability to manufacture steel pressure vessels and other heavy components is in place.

But there is another potential hitch, which is those 1,250 hectares of farmland. How can they be lit? There is a choice: Either we position them in greenhouses on the surface using natural sunlight, or we provide enough power to illuminate them artificially indoors or underground.

Putting the greenhouses indoors would simplify the city architecture, but how much power would be needed? The average solar flux at Earth's distance from the Sun is 1,300 watts per square meter. But considering the day-night cycle and the shifting angle of incidence throughout the day, atmospheric losses, and occasional clouds, this works out to a daily average received solar flux of about 250 W/m² at the equator. But plants can grow well as far north as Norway, where the average day-night flux is around 125 watts per square meter, roughly the same as it is in subtropical latitudes on Mars. Using LED light bulbs, which are about 70 percent efficient at turning electricity into light energy, it would take (125 W/m²)(10,000 m²/hectare) (1250 hectares)/0.7 efficiency = 2,232 MWe to light up all the farmland. That's more than four times as much as all other power uses combined.

So, unless technologies are introduced that can produce very large amounts of power at very low cost, an extensive network of surface greenhouses, with a total area about three times as large as the city itself, will need to be constructed.

Developing a Mars City

Elon Musk has proposed creating the first Mars city as a metropolis of one million people, populating it over a ten-year period by flying in a thousand Starships per year with each carrying a hundred colonists.

I do not believe this is a realistic scenario. In the first place, as I have already discussed, the material logistical requirements of even a fifty-thousand-person city are quite substantial. A million-person metropolis would need to draw vast amounts of materials from a large array of facilities scattered over extensive distances. This won't be possible until there are railroads on Mars, which can't be built until Mars has a large-scale steel production industry.

It was 250 years after English colonists arrived at Jamestown before the United States sported a city of one million people, and that city, New York, was only able reach that level in 1860 based on a very extensive inland transportation network of road, river, canal, and rail. Million-person Mars cities are quite a way off.

But even fifty-thousand-person Mars cities won't be created using methods comparable to the D-Day landing at Normandy during World War II, because the logistical difficulties of supplying fifty thousand people on Mars from Earth are much greater than those involved in supplying 133,000 troops on the beaches of Normandy from bases in Great Britain. Fifty thousand people consume a hundred tons of food *per day*. Even a small World War II Liberty ship could pack ten thousand tons of cargo and cross the English Channel in a couple of hours. But Starships only carry a hundred tons each and would take more than two years to do one round trip across interplanetary space. Shipping a hundred tons of food per day from Earth to Mars would be unthinkable. The only practical way to feed fifty thousand people on Mars is to grow the food on Mars. The greenhouses to do this need to be built before most of the people arrive.

Instead of being accomplished by a massive Normandy beach–type landing, Mars cities will be created by a process of growth more comparable to that which led to the eventual development of towns and then large cities in America and Australia.

Explorers will first search the land to find the best site for the city, and then one or two Starships will land there to establish a fifty-to-one-hundred-person-scale Mars base. Those first pioneers will then go to work, building greenhouses and other production facilities, as well as habitation domes and tunneling to create still more habitable volume underground. Once these are operational, it will be possible to bring in another hundred settlers, enhancing the local skill pool and labor power, thereby accelerating the construction of still more habitation and resource production facilities. In this way, the settlement might double in size every two years (launch windows from Earth to Mars occur every twenty-six months), until it reaches several thousand people about a decade after it is founded. But immigration of a thousand people per year requires twenty Starships arriving during every biannual launch window, which, counting tankers, in turn requires a hundred Starship launches from Earth per year to support. While *total* Starship-type vehicle launch rates from the entire planet of Earth might eventually be well more than that, it is unlikely that any one city would receive logistical support of a much higher order.

Based on such immigration rates, the city might reach a population of fifty thousand in half a century. However, if the population was highly fertile, it could grow significantly faster. For example, if the city population were split evenly between men and women, and the average immigrant woman had two children after arriving on Mars, the effective immigration rate would double. Over time, native-born children would become a more significant contributor to population growth than immigration. For example, if the average girl born on Mars has four children in her lifetime, the city would have a birth rate of about twenty-five per thousand per year, matching immigration once the city reached forty thousand people, and doubling the population roughly every thirty years thereafter.

Taking all this into account, we could see fifty-thousand-person Mars cities within thirty or so years of the beginning of settlement, reaching a hundred thousand in sixty years, and growing to a million-person metropolis about a century after that.

To be clear, I am *not* saying that in the ideal Martian city, the average woman will have four children or anything of the sort. All I am saying is that, beyond a certain size, a city's growth rate will depend primarily upon its birth rate. Some cities will not develop a culture based on large families and remain small. But the ones that do will outgrow the rest.

Our fifty-thousand-person Mars city-state can therefore be considered either as a snapshot taken during a phase of the growth of a future Mars metropolis or as a vision of a way of life that may continue in numerous smaller Mars cities for centuries to come.

But how can such cities be built?

Elements of City Construction

Initial bases on Mars can be started simply by landing several habitable spacecraft close by, and then linking them together with surface tubes or subsurface tunnels. But the first requirement to create a real city is to make a pressurized volume large enough to contain the homes of thousands of people. There are several proposals for how to do this. These include building domes on the Martian surface, tunneling underground, into cliff faces, or glaciers, or making use of natural structures including lava tubes, linear canyons, or ice-filled craters. Let's consider the merits offered by each of these.

Domes

Science fiction has frequently featured extraterrestrial cities protected from outside vacuum conditions by large domes. Such designs are possible but have limits.

The principal limit to the size of a domed Martian city is the strength required of the dome materials. In order not to burst, the ratio of the diameter of a hemispherical dome D to the thickness of its material t must be less than twice the ratio of its yield strength σ to its pressure P. In equation form, this can be written:

$$D/t < 2\sigma/P \quad (7.1)$$

If the shape is a cylinder instead of a dome, the diameter must be half that shown in equation 7.1 above. In practice, even for a sphere, we would want the diameter to be somewhat less than the maximum given by equation 7.1 to provide a safety factor.

For a given amount of material and dome thickness, the maximum possible dome diameter is inversely proportional to the city's pressure. Earth's sea level pressure, also known as 1 bar, or 1,000 mb, is 14.7 psi. But we don't need to pressurize our city to that degree. Where I live, near Denver, the air pressure is 12.5 psi (850 mb), and everyone here gets by just fine. People who climb Mount Everest must deal with 5 psi. This is a problem if the mountaineer is breathing air, because air is only 20 percent oxygen. Air at 5 psi only includes 1 psi of oxygen, and that is too little to support healthy respiration. Mountain climbers get around this by carrying oxygen tanks, allowing them to breathe 5 psi of oxygen atop Everest, which is more than the 3 psi of oxygen contained in sea level air. Apollo astronauts also breathed 3 psi oxygen while they were in their spacesuits. This was a good choice for them because the lower the pressure of the spacesuit, the lighter and more flexible it can be. But pure oxygen, even at reduced pressures, can be a fire hazard. So, on the Skylab space station, NASA provided astronauts with a 5 psi (340 mb) atmosphere, including 3 psi of oxygen and 2 psi of nitrogen. This worked well. It would also be a good choice for a Mars city, as it allows the city to operate at reduced pressure, sharply cuts nitrogen requirements (nitrogen is only 2.6 percent of the Mars atmosphere, and so involves a lot of effort to extract), and would also allow astronauts to switch to 3 psi pure oxygen-fed spacesuits without risking decompression sickness ("the bends").

Therefore, let's choose a Skylab-type 5 psi (3 psi oxygen, 2 psi nitrogen) atmosphere for our city. The strongest material currently known is Spectra, a polymer similar to the better-known Kevlar, which has a yield strength of 500,000 psi.

Plugging 5 psi for our pressure and 500,000 psi for the yield strength of our material, equation 7.1 reduces to simply D/t < 200,000. So, that means, if we use choose to make our dome out of 1-mm-thick Spectra (either a Spectra net with an average thickness of 1 mm or similar materials currently in development that are transparent[4]), the maximum diameter our dome can have, with no safety factor, would be 200 meters. To get a safety factor of 2, we could either cut the diameter in half or double the thickness of the walls. Either would be feasible. But if we want to get significantly bigger, the walls need to be much thicker. Assuming a safety factor of 2, a 600-meter-diameter dome would need to be made of 6-mm-thick material. A 180-meter-diameter dome made of 1.8-mm-thick material would have a mass of about 90 tons. The 600-meter-diameter dome would have a mass of 3,400 tons. It would take ten of the 180-meter domes to create a surface area roughly equal to that provided by the single 600-meter dome, but even so, their total material mass would only be 900 tons, or one-quarter that of the 600-meter-diameter unit.

Moreover, if we had to transport the domes from Earth, we would be limited by the Starship's hundred-ton payload capacity. At ninety tons mass, the 180-meter-diameter dome could thus be supplied premade from Earth, a major advantage for initial settlement. Each such unit would enclose 2.5 hectares of land. Larger domes would need to be manufactured on Mars.

The domes would need to be anchored to the ground. This could be done either by excavating hemispherical craters and then filling them in with regolith or water, or, more economically, by spiking the edges of the dome deep into the regolith. The latter could be done using perforated, barbed tubes for spikes and running hot water down through them after they have been driven into the ground. The water would spread out into the surrounding regolith and then freeze, cementing the tubes firmly into the resulting, extremely strong permafrost. Either way, however, anchoring ten 180-meter domes would be a much easier job than dealing with the 600-meter dome.

So, if we are going to employ domes as the basis of Mars's cities, we will need to do it by networking large numbers of moderate-sized domes, rather than single huge ones. The networked dome approach also provides safety advantages, since if any one of the domes were to experience a catastrophic failure, its residents could readily move into the other ones.

Instead of using domes, we could use linear cylindrical sections. These would have to be twice as thick as domes for a given diameter and pressure, but they could be of any length. So, if we used material with the same 1.8 mm material proposed for the dome, we would be limited to a cylindrical form 90 meters in diameter. But using the same 90 tons of material, we could cover a region 360 meters long. This would enclose 3.2 hectares, making it more efficient than a dome. Moreover, it could be extended in length by being constructed out of several sections brought by several Starships. Width could also be expanded indefinitely, without the need or cost of having a huge radius of curvature structures, by positioning such units side to side, with strong Spectra tethers attached to their joined interior edges anchoring them firmly into the ground (see Figure 7.14).

Such structural techniques could thereby be used to provide much larger single volume areas to the city's settlers or create linear systems with lines of sight that could extend for kilometers. With the right design, wind could be generated within these large volumes, bringing fresh breezes scented by indoor orchards to all the living areas of the city.

Technically speaking, cylindrically covered pressurized habitation volumes might be called "roofed" structures. But most of what I'm going to say about domes refers equally well to structures of this type. So, to avoid getting too wordy, in what follows, I'll just use the term "domes" for both linear and round surface pressurized systems of this general type.

To preserve heat and prevent water frosting out on the inner surface of the pressurized domes, each would be surrounded by a slightly larger, unpressurized, transparent dome (say a 181-meter outer dome diameter for a 180-meter inner dome). Since it would not need to contain pressure, this outer dome could be very lightweight. The ~50 cm gap between these two domes would contain Mars air at Mars pressures, effectively insulating the inner dome from the cold Martian wind.

The advantage of domed cities is that they are on the surface, illuminated with sunlight. They could therefore be filled with conventional houses—perhaps omitting roofs that may be unnecessary under smaller domes—with tree-lined walkways or bike paths, and possibly fish and swimming ponds as well. While farming can in principle be done using artificial lighting, the power burden associated with such techniques is extremely large. Domed habitation areas could provide the acreage for farms lit by natural sunlight. Rows of narrow-diameter, long-tube, cylindrical-roofed greenhouses can do this job as well, with less dome mass and less excavation required per useful acre of pressurized, sunlit land. They lack the aesthetic quality and recreational possibilities of large-diameter open field farms, but since a thin-walled ten-meter-diameter greenhouse tube with ten times the length can cover the same area as a shorter thick-walled hundred-meter diameter one using one-tenth the dome material mass, the increase in efficiency gained by going narrow for greenhouses is something to be reckoned with. This is especially so since, as we have seen, a Mars city will require three times the farmland to grow its food as the area comprising the town itself. Martians will probably make use of both.

One objection to living in domed cities, however, is cosmic radiation. The Martian atmosphere has a mass density looking straight up from the surface to the zenith of about 16 gm/cm^2, which is more than that required for a solar flare storm shelter. But most solar flare particles would not reach the surface coming straight down. They more typically enter the atmosphere coming in on a slant, which quadruples the average shielding it provides to an average of 65 gm/cm^2. As a result, despite Mars's lack of a magnetic field, a person on the surface of Mars would be at least as well shielded against solar flares as astronauts on the International Space Station, who enjoy solar flare protection from the magnetic field of the Earth. But like an ISS crew member, a colonist living on the surface of Mars would be exposed to galactic cosmic rays at about half the intensity experienced by an astronaut cruising through interplanetary space. (It's half because she would have Mars beneath her, which—as Earth does for ISS—blocks out half the sky.) This dose rate would be on the order of 15 rem (0.15 Sievert) per year. If one accepts the extremely conservative, linear no-threshold (LNT) method[5] of assessing radiation risk, this would increase the risk of developing a fatal cancer sometime later in life by about 1 percent for every five years spent living on the Martian surface.

This is not a showstopper. The average American smoker increases his cancer risk by 20 percent. A Martian living full-time on the surface for half a century would only accumulate an increased risk of 10 percent. (That's the number you get if you believe in LNT methodology, which greatly exaggerates the risk from low-dose-rate radiation exposure. According to LNT methodology, if drinking a hundred glasses of wine in one night would kill you, then drinking one glass would expose you to a 1 percent risk of death. That is false.) But still, it would be nice to reduce this risk. That can be done by living in houses that have much of their living quarters underground and placing most workplaces underground as well. A few meters of Mars rock, soil, or ice are all it takes to completely shield out galactic cosmic rays. If a Martian were only to spend six hours per day outdoors or in the aboveground portion of his house, his dose rate would be cut by a factor of four.

Even sharper reductions can be achieved by building the city underground.

The Mars Underground

Many Earth cities today include subway systems that are big enough to house cities themselves. There is no reason comparable networks of artificial caverns could not be built on Mars.

They would have to be spruced up significantly to make them attractive places to live, as in most cases the aesthetic standards of subway stations are pretty low. But higher standards are available, with readily available examples being the indoor environments of many upscale shopping malls or the classier terminals some airports reserve for international flights. (Note to readers: There is a secret world of high-class international airport terminals unknown to travelers who only fly on domestic flights. Despite flying frequently from Denver to Los Angeles on business, I was unaware of this for years, because I only went through the domestic flight terminal at LAX, which, as everyone knows, is a dump. But then I took a transfer through the LAX international terminal to Australia and saw the secret city. Wow.)

FIGURE 7.1. *Underground structures can be built with attractive aesthetics. In the Twardowsky Mars colony design developed by a Polish team, the city is built into the side of a crater, providing natural sunlight with very little radiation flux reaching the interior.*[6] *(Credit: Wojciech Fikus, Wroclaw University of Science and Technology)*

Most large shopping malls and airport terminals are built aboveground and do not need to support a heavy load of cosmic ray–shielding material above their roofs. But this additional requirement can readily be accommodated on Mars.

The 5-psi atmosphere we have baselined for our atmosphere exerts a force of 34,000 Newtons per square meter. Since Mars's gravitational acceleration is 3.8 m/s², that means a pressurized structure with zero structural strength would be able to suspend about nine tons of material per square meter on top of its flat roof. That's more shielding than Americans enjoy across most of the Rocky Mountain West (in Denver, we have about 8.5 tons per square meter). Mars's regolith has a mass density of about five tons per cubic meter. So, a layer of soil and rock about two meters deep would be sufficient to do the job. Alternatively, nine meters of water could be used. In this case, the indoor mall-like city building would essentially have an artificial lake atop its roof. This could be shielded from Martian cold by a thin, lightweight, no-pressure plastic dome of the type discussed above for temperature screening of domed cities. Alternatively, the top of the water could be allowed to freeze over, creating an ice-covered lake like those found in America's northern states in winter. Either way, the lake could be stocked with aquatic plants, which would oxygenate the water, and then fish, turning it into a combination city skylight and rooftop aquaponic garden.

If the city is positioned under a pressurized dome (i.e., if there were an underground or underwater expansion to a domed surface city), the same approach could be taken simply by pressurizing the underground portion higher than the surface 5 psi. For example, we could pressurize the underground portion to 6.5 psi. This would be enough to support a layer of water 2.7 meters (~9 ft) deep under the domed-surface zone, creating a nice fish pond and swimming pool for the region above and ample shielding to the habitats below.

Such an approach would require airlocks between the underground quarters and the domed-surface region. If this inconvenience is to be avoided, the underground system will need to be built with structural strength of its own. This could be done using steel, which can be readily manufactured on Mars out of the rust-red iron oxide that adorns the planet almost everywhere. But such an advanced material is not necessary for this purpose. Truly grand ancient and medieval structures were built using little or no steel. Instead, the builders of the Parthenon, the Roman aqueducts, and Gothic cathedrals took advantage of the principle that very simple materials, such as stone and brick, can nevertheless be extremely strong under compression.[7] Roman vaults or Gothic cathedral arches can support tremendous loads—and have done so on Earth for centuries, despite the blows of weather and human violence, carrying burdens greatly in excess of that needed to shield an underground city on Mars. Taking advantage of modern construction machinery, large high-ceilinged,

Gothic-cathedral-type underground pressurized spaces could be created rapidly on Mars.

Alternatively, the underground city could be built by tunneling through hard rock that would supply its own structural strength. There are tunneling machines called roadheaders that can do this very efficiently. For example, the Sandvik company currently manufactures the MT361 roadheader, which runs on two hundred kilowatts of electrical power and can tunnel at a rate of about 500 cubic meters per day. Its total mass is 60 tons, making it transportable to Mars. It can cut tunnels 3 to 5 meters high and 5 to 7 meters wide. So, setting it at 5 meters high and 7 meters wide, it could advance into bedrock cutting a tunnel at a rate of 14.3 meters per day, or 5.2 kilometers per year. 5-meter height is sufficient to allow two floors for a habitation, each with 2.5 meters (8.3 feet) of headroom. So, let's say we keep 3 meters of the 7-meter width for a street—we would have 4 meters of depth left for housing. A two-story townhouse 4 meters deep by 20 meters long would have 160 square meters of floor space, or 1,600 square feet, making it quite suitable for a family of four. There would be room for 260 such houses along the 5.2-kilometer tunnel, providing sufficient housing for more than a thousand people. If we wanted to create habitable space faster, we could simply employ more machines.

Martian cities based on row housing might seem unappealing. But the skillful city architect could come up with much more interesting patterns of winding tunnel streets intersecting perhaps with spacious domes and hillside promenades offering broad views of the surrounding countryside. As Canadian engineer and Mars city designer Paul Meillon brilliantly put it: "Michelangelo famously said, 'Every block of stone has a statue inside of it, it is the role of the sculptor to discover it.' It falls upon our shoulders to discover the hidden city lurking within the Martian bedrock."[8]

By building such underground cities into hills, mountains, mesas, or canyon walls, sides can be left open to be covered by large windows, thereby providing spectacular views of the surrounding landscape from the city's public areas.

FIGURE 7.2. *Hillside locations could provide nice views from many areas of a Mars city. Here we see the hydroponic garden area of the Twardowsky Mars City design. (Credit: Wojciech Fikus, Wroclaw University of Science and Technology)*[6]

Much vaster pressurized volumes can be obtained by taking advantage of natural features, including lava tubes, glaciers, linear canyons, and ice-covered craters.

Linear canyons, called fossae, are plentiful on Mars and can extend for kilometers. The architect Stefanie Schur[9] has proposed that these be roofed over with cylindrical transparent roofs of the type we discussed earlier to provide attractive locations for Mars cities. The volume enclosed would be much greater than what you can obtain by roofing over a flat plain. More important, the surface area beneath the transparent roof would not only be larger, but much more interesting, as it would include not merely a flat plain, but two hillsides facing each other. These hillsides could be covered with houses and shops, giving the Mars city a look remarkably like many charming, steep hillside Mediterranean towns such as Sedona, Mykonos, and Cinque Terre. The use of Martian materials would lend itself to Mediterranean, Moorish, and Santa Fe architectural styles as well. Schur also notes that building the colony along steep canyon sides would tend to promote good exercise, since people would have to frequently climb up the slopes to get where they wished to go. Nestled between two surrounding hillsides, thus blocking cosmic rays from most directions, the city residents could enjoy plenty of time outdoors with radiation doses cut well below dangerous levels.

FIGURE 7.3. *Cinque Terre, Italy, is built along steep hillsides. Mars cities using similar architectural and city design styles could be built within roofed-over fossae canyons on Mars. (Credit: Wikipedia Commons)*

Lava tubes form when flowing lava cools and solidifies at the surface while continuing to flow hot underground.[10] When the flow eventually stops, a large empty tube can be left behind. Many lava tubes are known on Earth, and they have also been discovered by remote sensing on the Moon and Mars. Terrestrial lava tubes, such as those in Hawaii, are frequently the size of subway tunnels. However, because of the lower gravity of the Moon and Mars, lava tubes on those worlds can be much larger. In 2014, a group of middle school students from Evergreen Middle School in Cottonwood, California, studying images taken by NASA's Mars *Odyssey* orbiter, discovered a Martian lava tube with an entrance diameter of 180 meters.[11] It is believed that many much larger tubes, with diameters in the 250-meter to 500-meter range and lengths of tens of kilometers are waiting to be found (and will be, once NASA sends an orbiter with the right kind of ground-penetrating radar). If fitted out with an airlock-equipped entrance, such tubes could be pressurized to provide enormous living spaces for Mars colonists.

Lava tubes and other types of natural caves can only be used where you find them. But there are regions on Mars where large caverns can be created fairly easily at will. These are the giant glaciers of the planet's northern hemisphere. Extending southward as far as 38 degrees N (the latitude of San Francisco on Earth) and covered by only a meter or so of dust, these vast ice formations contain more water than America's Great Lakes. Mars colonists wishing for a cavern to house their city could make one to order just by melting it out.

For example, let's say they wanted a cavern in the shape of a half-cylinder, with a flat bottom and a round top, and a side-to-side diameter of fifty meters. They bring with them a set of nuclear reactors with a total power of a thousand megawatts of electricity, roughly the amount used to power a city the size of Denver. In addition to electricity, these reactors also generate about two thousand megawatts of waste heat, producing a combined three thousand megawatts, all of which can be used to melt ice. With this much power available, the colonists could melt ice at a rate of thirty metric tons per hour, or about 320,000 cubic meters per year. This would allow them to tunnel into the glacier at a rate of 325 meters per year, creating living space as they go. Within a decade, they would create an ice cavern more than two miles long. Lined with waterproof material to insulate the floor and the ceiling, the entire enormous volume would become pressurizable, habitable space. A large underground liquid water lake could also be created. Lit up with LEDs powered by a nuclear reactor, this could become a vast fish and kelp farm to provision the growing colony.

Waste heat from a nuclear reactor could also be used to transform ice-filled craters on Mars into ice-covered lakes.[12] The lake would be lit by natural sunlight, and thus become a habitable environment—making its creation the first step in terraforming Mars. Large inflatable dome-like structures with open bottoms and raft-like floors could be placed within the lake to house the city, whose inhabitants could enjoy fishing and SCUBA diving, among other recreations.

One could go on discussing further engineering options for creating the pressurized volumes for Martian cities. My favorite at the moment is to have the city primarily aboveground, with its pressure barrier configured as a large disk with a nine-meter-deep lake as its skylighted roof. The water would provide optically transparent radiation shielding and a home for fish and aquatic plants. At the same time, its weight would balance the gas pressure from below, leaving only the cylindrical section walls with a requirement to withstand pressure. The need for a deep regolith anchoring system would be radically minimized as well. Assuming the cylindrical section were 20 meters high, a 500-meter-diameter cylinder wall would only need 31,400 square meters of pressurized Spectra, less than 1/12 of the 392,500 square meters required to build a 500-meter-diameter hemispheric dome. Beneath the disk's 20-meter-tall (66 feet) ceiling, a town could be built, and networks of similar structures could be expanded to grow the city. Suffice to say there are many ways to build Martian cities that are both sound and functional,[13, 14] or as Vitruvius put it, "firm and commodious."

But Vitruvius set a third requirement as well. Cities must be built to be "delightful."

How can that be accomplished on Mars?

Making Martian Cities Delightful

Martian outposts don't need to be beautiful, or otherwise enjoyable. They just need to be functional. But Martian cities must be delightful. This is so because, in order to come into existence, Mars cities must attract immigrants. Those that do will grow. Those that do not will be stillborn.

Not long ago I was visited by a Chinese visionary who showed me his design for a Mars city. It consisted of a single tower three hundred stories high. From an engineering point of view, the concept was arguably plausible. Mars's lower gravity allows structures to be built taller, and his concept eliminated the need for domes or vaulted undergrounds. Keeping the building diameter modest also eases both the structural problem of holding in pressure and anchoring the building to the ground. But from the social point of view, I think the concept makes no sense. Who would want to live in a city comprised of such a structure?

Some aspects of Martian city aesthetics are predictable based on objective conditions. For example, it is far more likely that Martian houses will be built of adobe, brick, or stone rather than wood, because the former materials will be easy to come by or make on Mars, while the latter will need to be grown with considerable effort. While Martians might use wood to create their more elegant furniture, it will be quite some time before it is available in the quantities required to make much use of it as a homebuilding structural material. So, for the same reason that they prevailed in their characteristic regions on Earth, Santa Fe– or Mediterranean–style buildings may well prevail on Mars.

As to urban plans, that is a matter of taste, over which architects and Mars city planners will no doubt differ greatly. For myself, I agree with Jane Jacobs[15] that the most charming cities are those that are unplanned.

Jacobs was the antagonist of the famous New York City planner Robert Moses. In the postwar period, Moses sought to rebuild New York by tearing down old neighborhoods and replacing them with modern residential high-rises spaced out amid small urban parks, with retail stores, restaurants, offices, repair shops, factories, and playing fields all zoned elsewhere in their carefully planned proper places. Jacobs's ideal, on the other hand, was the old-style New York neighborhood, with low-rise residences, stores, bars, and sweatshops all mixed helter-skelter, creating lively foot traffic and a vibrant street scene whose elderly observers, chatting or playing chess or dominoes on the tenement steps, also provided many watchful eyes over the kids playing stickball and hopscotch in the street.

Jacobs argued that Moses's high-rise projects would become unsafe, particularly at night, because there would be no foot traffic around them. She proved to be right. In contrast, the lively unplanned neighborhoods she fought to preserve were comparatively safe, and much more fun. Wandering amid their chaos, there was always a used bookstore, antique shop, a loft ballet school, or something else interesting to discover, or, as the song went, you could just enjoy "standing on the corner watching all the girls go by."

So, I'm with Jacobs. If it were up to me, Martian cities would be unplanned, or planned to be able to grow unplanned, in an even more chaotic fashion than the eclectic Lower East Side or Chinatown. Make them more like the delightful medieval-vintage sections of European cities, with a random street plan aboveground and below fostering a mix of homes with cafés, bakeries, peddler stalls, taverns, and second-story music schools, ready to be visited by pedestrians, cyclists, or roller skaters game to thread their way through the soccer balls being joyously kicked around by the kids in the street. Let's have a big central dome in the city center to create a park for massive civic events, and perhaps house some iconic buildings, too. But for the rest, let life discover the city hidden within the bedrock.

That would be my choice. But it won't be up to me. Martian architects will make their own decisions. Some might do it my way, others might use the characteristic modern rectangular city block grid, a radial plan like Pierre Charles L'Enfant laid out for Washington, DC, in 1791, or even the three-hundred-story skyscraper city plan proposed by my Chinese friend. But not even the Martian architects will be able to make the final call as to which plan gets used. No, the last word on Mars city design will be given by the immigrants.

Reason proposes, reality disposes. The evolution of Martian cities, like that of biological species on Earth, will be governed by natural selection. The cities that attract the most immigrants will grow. Those that fail to do so will disappear. The city architectures that prevail will be the ones that people like the most.

I'd say, "may the best plans win," but that would be redundant. The ones that win will be the best.

FOCUS SECTION

A GALLERY OF MARS

CITY DESIGNS

Many concepts have been advanced for the architecture of Mars cities. Here's a gallery of some of those I like the best. Many more can be found in the *Mars Colonies* and *Mars City States* books published by the Mars Society.

FIGURE 7.4. *Large domes planted with fruit trees can be used to create central parks for Mars cities. (Credit: Bryan Versteeg / Spacehabs.com)*

FIGURE 7.5. *Narrow cylindrical structures are best for the greenhouses that will produce most of the city's crops, as they require less material per hectare.*[16] *(Credit: Bryan Versteeg / Spacehabs.com)*

FIGURE 7.6. *The American architect Stefanie Schur proposes roofing over fossae canyons to create long linear cities. Housing and shops would line the steep canyon walls in much the same way as they do in many charming hillside Mediterranean towns.*[9] *(Credit: Stefanie Schur)*

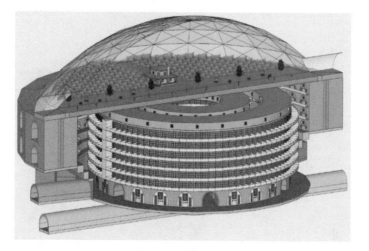

FIGURE 7.7. *The Australian Southern Cross Innovations group recommends using steel and concrete to create coliseum-type buildings with domed parks on each roof. Somewhat similar buildings, known as Gasometers, can be viewed in Vienna.*[18] *(Credit: Southern Cross Innovations)*

FIGURE 7.8. *Greek American engineers George and Alexander Lordos propose building their Star City, Mars by tunneling into a hill, adorning its tunnels with numerous domed outdoor parks.*[19] *(Credit: George and Alexander Logos)*

FIGURE 7.9. *An interior view of the Lordos brothers' Star City on a hill. (Credit: George and Alexander Lordos)*

FIGURE 7.10. *Named after a legendary Polish space traveler, Twardowsky Colony is built into the sloping sides of Jezero Crater, the landing site of the* Perseverance *rover.[6] (Credit: Wojciech Fikus, Wroclaw University of Science and Technology)*

FIGURE 7.11. *Australian engineer Muhammad Akbar Hussain recommends a Mars city design based on a network of domes connected by both aboveground and subsurface tunnels. A hexagonal network of strong Spectra cables reinforces the domes with windows of weaker but optimally transparent silicone material filling in the gaps.*[20] *(Credit: Muhammad Akbar Hussain)*

FIGURE 7.12. *The Spanish Sustainable Offworld Network has developed its Nuwa City architectural design in detail. The city employs tunnels in the side of a cliff to provide radiation-shielded habitation, with sweeping views of the land below from the cliffside windows.*[21] *(Credit: ABIDOO Studio / SONet, Sustainable Offworld Network)*

FIGURE 7.13. *Canadian engineer Paul Meillon proposes creative tunneling to "discover the hidden city lurking within the Martian bedrock."*[8] *(Credit: Paul Meilllon)*

FIGURE 7.14. *The French Team Foundation proposes the use of cylindrical dome sections to create vast pressurized spaces illuminated by natural sunlight. The lines where the cylinders meet are anchored in the ground by strong tether cables.*[22] *The structure thus bears a striking resemblance to a Gothic cathedral, but with all elements in tension rather than compression. (Credit: Michel Lamontagne)*

FIGURE 7.15. *An Australian American team has proposed building its city in the ice-filled Korolev Crater. Waste heat from the city and its power reactors will turn the crater into an ice-covered lake. This could enable the creation of an extensive local aquatic biosphere, including fish and kelp, and provide settlers with recreational opportunities offered by an indoor beach.*[23] *(Credit: Epifano Pereira)*

FIGURE 7.16. *The international Nexus Aurora team proposes covering its buildings and plazas with rooftop fish farms, providing both natural sunlight and ample shielding for all below.*[24] *(Credit: Nexus Aurora)*

8

SOCIAL CUSTOMS
ON MARS

In hostile environments, only groups with institutions that promote extensive cooperation and sharing can survive at all.

—JOHN HEINRICHS, *The WEIRDEST People in the World*[1]

And now that I have come near the end of this book in which I have recorded so many considerable achievements of the Americans, if anyone asks me what I think is the chief cause of the extraordinary prosperity and growing power of this nation, I should answer that it is due to the superiority of their women.

—JOSEPH HENRICH, *Democracy in America*, 1835[2]

WHAT WILL MARTIAN SOCIETIES BE LIKE? What will be their mores, customs, and characters?

These questions are worth investigating, not so much to make recommendations to future colonists, who will, while no doubt reading this book with great interest, only adopt such ideas with which they already agree, but rather to understand what it is they are likely to create.

In this chapter, therefore, I will attempt to divine the probable shape and customs of future Martian societies, not necessarily my own preferences, but rather what I think are likely to be most successful, and therefore prevail as the dominant forms in the end. Many of these may contradict ideals currently upheld by many on Earth, or prominently include features of social life, such as organized sports or religion, in which I personally have little interest. But be that as it may, the evolution of Martian societies will not be determined by what I or others alive today want, or even what the early Mars colonists themselves may want, but by what proves to work the best.

The Martian cities whose cultures set the mold for the Red Planet will be those whose populations outgrow the rest. People will arrive on Mars in only two forms: either as immigrants or as children. While immigrants will comprise most new arrivals early on, the city-states where they prove fruitful and multiply will be the ones that take off.

In short, Mars needs women.

The importance of female immigration is illustrated strikingly by the comparative fortunes of the British and French colonies in North America. The British colonies were created primarily by people interested in setting up new communities, with all that they imply. Consequently, British women were willing to emigrate there. In contrast, the French viewed North America as a region for the establishment of profitable fur trading posts, and thus few women went. As a result, even though the population of France was four times that of Britain, by the 1750s, when the struggle for control of North America was decided, there were two million English-speaking people on the continent, and fifty thousand French. The result was inevitable.

Mars city-states will need women, and not just any women, but ones interested in having children and raising families. And they will need men willing to help them do so.

So, I think that currently fashionable ideas about the purpose of romantic relationships being a form of self-realization, recreation, or game, will have to be discarded. Instead, courtship will have to be viewed in much more traditional terms as part of a process of finding the right partner for a lifetime project of raising a family. This commitment will be taken seriously. Divorce will probably be legal, but it will be frowned upon, with those causing it being viewed as deserters or disruptors of something sacred and penalized by law or custom accordingly.

Mars also needs talent, actualized and unleashed. This can be obtained through immigration of skilled people and through education of the native born. But here we run into a difficulty because, while Mars will need women willing to give birth to and raise substantial families, it also will need their talents manifested across all areas of the economy. The dominant reality of Martian cities will be a severe labor shortage, in both quantity and diversity of available skills. Excluding half the population from participation in the creative workforce would preclude any city that chose such a path from being one of the winners.

This conundrum—how to reconcile career with family—fills many pages of women-oriented literature on Earth today. So far, modern terrestrial societies are not doing a good job of addressing it, forcing too many women to choose between them. The result has been a potentially catastrophic population implosion in most advanced countries. This can hardly be otherwise, as the attraction of a professional career on the one hand and the deterrent prospect of unpleasant single-motherhood after probable breakup of a "long-term relationship" with an unreliable boy-man on the other sends a strong message to young women.

Some countries have attempted to counter this trend with financial incentives, including subsidies to mothers in France, free daycare in some European social democracies, and tax deductions for families with children in the United States. But the success of such initiatives has been modest, as the amounts on offer have been far too small to be of much interest to all outside the lowest income brackets. For an educated, professional woman, for whom the costs of childbirth include not only the direct expenses associated with the child, but large amounts of lost income due to time away from work, the numbers don't remotely add up.

I think successful Martian city-states are going to have to deal with this in two ways. One, as alluded to earlier, is to create a culture where marriage is taken much more seriously. The other is to provide much stronger financial incentives for women—especially educated, professional women—to have children.

As an example, a city might decide to award substantial lifelong tax breaks to women in proportion to the number of children they have. If the city's tax structure is a flat 25 percent income tax, a woman might have hers reduced by 5 percent for each child she has, up to the first four, thereby reducing her tax rate from 25 percent to 5 percent. This might not reward her much during her childbearing years, as her earnings then would be low, but it would pay off big-time afterward. As educated people tend to be able to defer gratification, such a deal might prove to be very attractive to many professional women.

In fact, it might be too attractive, as it could result in a society where women control all the money, and thus the companies and political system. This would no doubt encourage female immigration, but men might not like it at all, causing a loss of male immigration, or even emigration. So, maybe the best plan would be to make the plan both more equitable and more marriage-affirming by awarding the tax break equally to both members of a married couple raising children, or just favor the women (who do the hardest part, after all) a bit with a 60/40 split in their favor.

In this connection, it should be added that the marriage practice in the most successful Mars city-states will almost certainly be monogamy. While, historically speaking, most human civilizations have practiced polygamy, it will be impossible for a polygamous society to prevail on Mars. In the first place, polygamy degrades the status of women, and thus would be a deterrent to female immigration, which is the critical factor in population growth. But polygamy also degrades not only the status but the dignity of the large majority of men, who are left without female partners and thus the possibility of having children. As a result, such men are driven to further degrade themselves through nonconsensual sexual practices,[3] develop violent personalities, and, as they have no stake in the future of the community, very frequently become criminals. The effect of having a large population of wife-denied men can be seen both in the dangerous "street" of many Muslim countries, and especially strikingly in China. There, the one-child policy implemented by the Chinese Communist Party in 1980 caused many couples desperate for a son to either abort or otherwise dispose of female babies, resulting in a large surplus of boys over girls. As these boys came of age, the rates of murder and other violent crimes in China soared.[4, 5]

Given its vulnerability to destruction by sabotage, a Mars city can ill afford to host a large violent criminal subculture. So, polygamy is out. (Polyandry is out as well because it simply doesn't work. A woman needs a husband who will devote himself to her children. This only happens when he knows they are his.[6] There is yet another reason why large-scale practice of polygamy is impossible for a successful Mars colony, and that is economic. As discussed earlier, the economy of Martian city-states is likely to rest substantially on invention. Neither the enslaved females nor the dispossessed mass of degraded males in a polygamous society is likely to have much opportunity to participate in the process of invention, drastically cutting down the number of potential inventors to the few elite men. A highly inventive society needs to be much more egalitarian.

The imperative to invent also means that Mars cities will need to be free. Furthermore, they will need to encourage the immigration and development of free-thinking individualists, because that is what inventors are.

This requirement will be in tension with a Mars city's need for social solidarity.

While invention requires individualists, survival and success of a new settlement in a hostile environment require strong community solidarity. When we look at the three most remarkable examples of such settlements in modern history—the Puritan settlement of Massachusetts in the 1600s, the Mormon move to Utah in the mid-1800s, and the Jewish migration to Palestine in the twentieth century—we see that all three groups achieved such solidarity through strong religious motivations and community bonds.

There are two problems with this model, however, as it relates to Mars. In the first place, to survive, a Mars city-state will depend on its scientific virtuosity. Strong religious beliefs are not incompatible with scientific or engineering excellence. But theocracy is. One of the best engineers I have ever known was a Christian who believed that the world was created by God inside of a week some six thousand years ago, exactly as reported in the book of Genesis. This did not stop him from playing a central role in assuring the success of the 1976 Viking landings on Mars. But, based on his biblical information, he thought that its life-detection mission was pointless, and had he or others of his persuasion been in charge, it never would have occurred. Fortunately, however, he lived in the United States, under laws that would not allow his church or any other religion to dictate scientific truths from ancient texts, or determine which scientific questions were worthy of investigation.

Religion has not always been opposed to science. Indeed, a very strong case can be made that monotheistic religion in general, and Christianity in particular, had a central role in giving birth to science. This is so because the entire scientific enterprise is based on the faith that there are rational laws underlying the structure of the universe. If these laws were enacted by a God who created the human mind in the image of his own, then these laws should appear rational, and thus discoverable, by human reason. As Johannes Kepler, the Renaissance astronomer who discovered the laws of planetary motion based on precisely such belief, put it: "Geometry is one and eternal, a reflection out of the mind of God. That mankind shares in it is one reason to call man the image of God." Similar passionate faith in the lawfulness of the universe has influenced many of the most profound scientific minds ever since, most notably Albert Einstein.

But while religion gave birth to science, it is not the same thing. It is true that some modern religions have adopted a non-fundamentalist approach that allows them to adapt to newfound scientific truths, thereby removing the overt impediment their beliefs might otherwise represent to scientific work. Yet, even in such cases, there is another problem in basing the community solidarity necessary for a Mars city-state upon a shared organized religion. It narrows the population from which the city can draw potential immigrants.

A Mars city-state based on a passionately shared common religion might well survive, if the religion was of the right type (i.e., open to newly discovered truths and agreeable to the idea of natural causality[7]), and even become prosperous. But it would be outgrown by those cities open to a broader array of immigrants. Compare Israel to the United States. Israel has been remarkably successful despite the notoriously politically fractious nature of its leaders and parties because of the extremely strong tribal solidarity its Jewish population can call upon in a pinch. But Israel is, and will remain, small. In contrast, by drawing immigrants from peoples all around the world and melting them into a new common identity, the United States became a superpower.

There may well be more than a few little Israel-type city-states on Mars. But the cities that outgrow the rest to become the basis of the dominant planetary civilization will be the ones that adopt something much more like the American melting pot model.

To be clear, the melting pot model is not the same thing as the multicultural models that have become fashionable lately in Europe.

FIGURE 8.1. E Pluribus Unum. *Immigrant children reciting the Pledge of Allegiance in a New York City classroom in 1892. For many years, the public school system has served as the frontline force for the melting pot of the United States. (Credit: Jacob Riis, Creative Commons)*

It is as open to immigrants from diverse backgrounds, but it seeks to change them into a new common identity. *E Pluribus Unum*—out of many, one. A Mars city-state would have great difficulty surviving if divided by competing tribal loyalties. It needs to be one people. But it can become one out of many.

A Mars city-state will need a common creed that can be subscribed to by all. In the United States, that was supplied by the doctrine contained in the opening lines of our Declaration of Independence. The Mars city will also need institutions that serve to spread its creed and meld its diverse immigrants together to form a strong new alloy. In America, historically, the frontline force of that essential melting pot has been the public school system.

In the United States today, the mediocre performance of much of the public school system and its progressive curriculum has driven some offended conservatives to propose its abandonment. As an alternative, they suggest that public subsidies should go to private schools with no common curriculum. While I agree that many of the problems they have identified are real and quite serious, I don't think their solutions are good. Indeed, they are merely the flip side of multiculturalism. Instead of rising to the challenge of maintaining an effective and unifying public school system, they have chosen to cut and run.

I don't think such an approach is likely to work well on Earth, and it would almost certainly be a disaster on Mars.

That is not to say the public school systems employed by successful Mars city-states will closely resemble those now dominant in the United States or other advanced Western countries. For one thing, both the Martian labor shortage and the generally more serious nature of the realities of life on Mars will require that kids grow up faster. Currently, we use our school system to keep young people out of the workforce and other adult responsibilities, extending adolescence far beyond its natural time. According to noted psychologist Bruno Bettelheim, it is this forced extension of adolescence that is responsible for the craziness that is characteristic of teenagers in America and similar countries today. Martian cities will need to raise more responsible youth. This can best be done by not segregating them into a separate children's world, as we do here now. So, my guess is that, starting with elementary school, school hours on Mars will only be about half as long as is typical on Earth, with students spending half their day working and associating with adults at home, in the greenhouse, in the shop, and in the lab—getting a good hands-on education in the process.

Secondary schools will thus not be run as reservations for teenagers, to keep them locked up and out of people's hair during daylight hours. Rather they will be strongly integrated with adult society, combining schoolwork with a series of internships.

The nature of the curriculum will also change. Instead of being oriented to organize students through drill and grill, schoolwork will feature a much stronger emphasis on developing scientific proficiency and creativity, with engineering design classes (in which students act either as individuals or in teams to design, and in some cases build, interesting systems) and technical sports playing a prominent role starting from the earliest grades. Young Martians will also learn history in their schools, because it is essential that children gain a common understanding of the values and grit it took to make their society possible.

A nation without heroes is a nation without a future.

Will successful Martian cities have universities? Some might follow the traditional, terrestrial model of higher education—after all, it has been shown to work. But perhaps there is room for radical innovation in this respect. Terrestrial universities do serve a useful function of conveying advanced knowledge—at least to some of their students. But they also serve a pernicious purpose of selling certificates—at considerable cost in time and money to the recipients—to separate those deemed learned from the unlearned. Such certificates have been made the necessary admission tickets to most higher paying jobs, and thus their procurement serves as a poll tax limiting the opportunities of a considerable portion of society. As a result, while some students do go to a university for the knowledge, most are there for the certificate. This degrades the university and imposes a huge waste of human labor and talent on society.

Perhaps a better plan would be to abolish the ivory tower and its certificates, and turn the entire city into a university. Technical subjects could be learned at work through on-the-job seminars. Love of literature could be promoted citywide through book clubs and public events and festivals like the Chautauqua talks that spread deep literacy through nineteenth-century America. Consider the thousands of frontier Americans who traveled for days to sit for three hours and listen to the Lincoln-Douglas debates, or the mass of artisans and farmers who provided the audience that Hamilton, Madison, and Jay addressed in *The Federalist Papers*. While we don't have standardized test results, there's plenty of anecdotal evidence to show that pre–G.I. Bill Americans—very few of whom went to college—were far more literate than the mass of today's liberal arts university graduates. My father, born in 1916, was an orphan who worked his way through high school selling newspapers on New York's Lower East Side. Yet he knew and loved his Shakespeare and derived much insight into human character from it. My uncle Alex, of the same generation, was a workingman, but he had an encyclopedic knowledge of history he obtained through voracious reading, and he put that understanding to work organizing several paint factories for the CIO (Congress of Industrial Organizations, a network of unions) in the 1930s. The fact that they were not exceptional is witnessed by the extraordinary efforts all sorts of public organizations took to deliver books to soldiers and sailors abroad during World War II.

FIGURE 8.2. *The soldiers and sailors of the Greatest Generation weren't typically college men, but they loved to read. American Red Cross volunteers collect books for the Victory Book Campaign in World War II. (Credit: US Army Center of Military History)*

You don't need to go to university to live the life of the mind. You just need to be part of a culture that is mentally alive.

The progress of Western civilization greatly accelerated in the regions affected by the Protestant Reformation. This was not because of any putative superiority of Protestant theology over the Catholic faith. Protestant theology is quite diverse, and on such key questions as free will verses determinism, the theology of the Church of England is virtually identical to that of the Roman church. No, the difference that mattered was that the Protestants translated the Bible into the local vernacular and encouraged their congregations to read it. This spread mass literacy and abolished the fundamental distinction between the learned and the unlearned. It was on this basis that the intellectual capacity of the people was mobilized.

The most successful Mars cities will create a vibrant popular intellectual culture. Perhaps the equivalent of a Protestant Reformation in education is in order.

Returning to the subject of social solidarity, I believe that successful Mars cities will need customs that reinforce it. Anthropologists studying diverse cultures on Earth have identified a range of such practices that generally serve this purpose. These include organized sports, group singing, moving together rhythmically to music, and participating in other group recreational, ritualistic, or practical (i.e., barn raising) activities directed toward a common goal. It may be observed that some combination of these has almost always been practiced in institutions, ranging from churches to the military, that seek to create team spirit among their members. While personally I am an individualist who has never had much affinity for organized sports, let alone organized religion, I think some activities along these lines will probably be necessary, and possibly quite prominent. I hope they won't be too all-encompassing, however, because while no Martian city (or any other society) could ever be successful if it were composed entirely, or mostly, of people like me, it will certainly need some.

Sports on Mars

Martian cities probably won't be able to spare the ground for sports that require huge areas of open land, like baseball, soccer, or football. But basketball, volleyball, and dodgeball are attractive possibilities that could be played in very exciting ways under Martian low-gravity conditions. More novel three-dimensional sports taking advantage of Mars's one-third gravity might well evolve, including perhaps something resembling the Quidditch games described in J. K. Rowling's delightful Harry Potter novels.

In addition to fostering team spirit, field sports on Earth serve the function of encouraging physical fitness. This will be even more important on Mars, where the low gravity will not impose the same involuntary workout that routine daily activities provide on Earth. But technical skills will be even more necessary for the survival of a Mars city than athletic abilities. Therefore, I believe that the most successful Mars city-states will incorporate technical sports like model aircraft racing and robot battling into both its schools (starting at first grade) and general culture. As educational exercises, such sports have the profound effect of reversing the typical relationship of students to knowledge. Instead of being a burden (as in "How much of this will we need to know for the test?"), knowledge becomes a weapon. Furthermore, while fostering solidarity in the same way that field sports do, technical sports also highlight the value of out-of-the-box thinking, and thus creative individualism, in a way that field sports simply don't.

Some sports, combining technical skill, courage, and endurance—for example, a Martian cross-country piloted rover race like the Alaska Iditarod—could readily become national festivals celebrating the virtues of the first Mars pioneers. Indeed, they could well become interplanetary events, with considerable income generated by selling the media rights on Earth.

Technical sports are occasionally practiced now on Earth with great success, but on a hit-or-miss basis, through robot competitions like those sponsored by First Robotics and the Mars Society. It would be a very positive development were such events made more central to school and civic culture. On Mars, I think they will be, with technical

FIGURE 8.3. *Alaskans still celebrate the grit of the state's first pioneers in the Iditarod cross-country dog sled race. A similar epic competition might well become a national festival on Mars. (Credit: Curtis Beach, Creative Commons)*

sports made a required class, and high school and neighborhood rover racing teams as closely followed and cheered by everyone as school and city football teams are in America today.

Dance and Music

When people move together in time with music, they tend to become psychologically bonded. This psychological phenomenon is utilized by diverse groups ranging from primitive tribes to modern armed forces seeking to create group solidarity and is also part of the "ice-breaking" fun of dancing with a partner. Community dancing is not a general practice in America today, but it once was. As a child I experienced it when my family spent some summers in a rural community that held Saturday night square dances. It really does work to bring people together. I believe that the most successful Mars cities will have similar customs.

Martians will dance.

All known terrestrial dance styles, past or present, would be possible on Mars. But certain types of dancing that involve movement in three dimensions would be greatly enhanced by Mars's lower gravity. The potential fun offered by these forms will be too wild to be missed.

Consider, for example, the possibilities offered by the Red Planet Jitterbug.

Martians are really going to swing.

My parents were both born in 1916. They told me that, in their youth, young people would frequently gather in parks in the evening and sing. That custom passed away before my time. But during my childhood, our family would sing together when we traveled by car. When I grew up and had my own family, however, my wife and I couldn't get the kids to sing along. I've asked around, and some families still do it, but there is no question that singing as a participatory activity has—except for church groups—largely passed out of our culture.[9]

Whatever the reason may be, we don't sing together anymore. I think this is a serious cultural loss that has contributed to the atomization of our society. Martian cities would do well to repair this deficiency, and I think the most successful ones will do so by creating a popular tonal musical culture that allows and encourages people to join their voices in song.

Martians are going to sing.

Sharing on Mars

Modern society is very inefficient in its use of capital equipment. For example, I live in an average suburban house equipped with its own oven, refrigerator, clothes washer, dryer, and dishwasher. My wife, Hope, and I also each have our own car. As far as I can see, all our neighbors are similarly equipped.

Collectively, this represents an enormous amount of material. Yet, except for the refrigerators, all the rest of these machines are typically in active use perhaps one single hour per day. For 95 percent of the time, they are sitting idle.

Well, I can afford to have my own household appliances. But what if, instead of buying an oven from Home Depot, I needed to have it shipped to me across interplanetary space? Even if the heavy steel parts of it were made locally, the electronics and other complex or fancy components might need to be imported at a cost beyond my means, or certainly beyond my desires. If I'm only going to be using the oven 5 percent of the time, wouldn't it make more sense to split its cost with a neighbor, or better yet, ten neighbors?

There was a time in my life when I had more limited means than I do now, and I did not have my own clothes washer and dryer. Instead, depending on where I lived, I either made use of the shared washer-dryer in the apartment building basement, or I patronized the laundromat down the street. Many city dwellers both here and world-wide still do exactly that. But in America, at least, nearly all families cook on their own range.

Perhaps Martians will do the same. But given the costs, I think there is a good chance Martians will choose a different path. Restaurants tend to be expensive, but a Mars city that embraced and promoted a culture of street food, with vendors selling dishes from stalls or carts, would provide real savings to its citizens, not to mention the fun, promising to make such towns lively places. Yet most people like to be able to choose what they eat and how it is cooked, and prefer to dine together at home as a family. So, I think many successful Mars city-states will feature community kitchens, where the ovens are always hot and you can bring your own ingredients, cook what you like, and then either eat it there or take it home for a private meal. Either way, the level of social interaction associated with cooking and eating is likely to be much higher than is typical now on Earth.

There will also be little reason for Martians to have their own cars. In the first place, Mars towns will be foot cities, where people mostly get around by walking or biking. To the extent a Martian occasionally needs the use of a car or truck, taking advantage of an hourly or daily rental from either a private company or the city government will make much more sense. So, there won't be the spectacle of untold thousands of stationary cars, made or imported at great expense, lining the city's streets doing nothing nearly all the time. In addition to saving on imports, the custom of vehicle sharing will thus make finding a parking spot on Mars much easier than it is currently on the home planet. (Many Earthlings might immigrate for that reason alone!)

Many other savings could readily be accomplished with other forms of equipment sharing. To make them viable, however, Mars city-states will need to have a patriotic culture that encourages responsible treatment of shared property.

So, to sum it up, at its heart the successful Martian city will need to have a vibrant culture, with a strong intellectual life, sports, music, dancing, social solidarity, sharing, and sense of purpose. This togetherness, originating from the necessities of frontier life, may ultimately prove to be Mars's greatest draw of all to citizens of Earth seeking something more meaningful than the lonely lives offered by the atomized society of the home planet.

Taxes on Mars

A Mars city-state will need to impose taxes because there are some activities that must be funded collectively. That said, it is important that the tax structure be as simple as possible to prevent the creation of a vast parasitical caste of tax accountants such as currently exists in America. Many of those involved in this racket are quite talented, and their relegation to an activity that produces nothing whatsoever of value to society is a massive waste of social capital that a Mars city would ill be able to afford.

Furthermore, an indecipherable tax code creates a spiderweb that can trap well-meaning citizens into offenses, turning them into criminals, with even greater loss to society consequently.

There are two types of taxes that seem most appropriate. One, previously mentioned, would be an income tax, with simple but strong deductions for those raising children—or who have raised children. The other is a tax on land, with its assessment fixed at its undeveloped value. This would encourage real estate development.

In contrast, sales taxes—which require some form of surveillance of all private transactions for their effective enforcement—are a huge waste of social time and an invitation to under-the-table activity. If personal income is taxed, taxing businesses as well seems counterproductive, since reinvestment of profits in business expansion would maximize economic growth.

The Practice of Law

The first thing we do, let's kill all the lawyers.

—SHAKESPEARE

American society today is inflicted with an enormous legal overhead. A significant fraction of our most talented young people are recruited to join the legal profession. They then spend the rest of their lives looting society by collecting huge fees to help people avoid ensnarement in the spiderweb of inscrutable laws and regulations they have spun. This has reduced much of the legal system to a protection racket, which together with related rent-collection swindles, such as those performed by income tax accountants, imposes a fantastic drain of social and economic resources that the Mars city-state would find extremely difficult to sustain.

But it gets worse because many of those who fall afoul of this apparatus must then be imprisoned, and the legal racketeers require that society pay the cost of that, too. Once again, given the extreme labor shortage faced by Martian cities, neither the loss of talent represented by large numbers of imprisoned people nor that of their guards would be acceptable.

Clearly, Mars cities will need a much simpler legal system. The fundamental principle for this already exists.

English common law contains a tradition, known as "mens rea"— Latin for "a guilty mind"—which is intended to protect people from prosecution for unintentional offenses. According to the doctrine of mens rea, if you did not knowingly commit a crime, then you are not a criminal, and should not be treated like one.

Every day brings us stories of citizens being prosecuted for actions no reasonable person could have suspected were crimes. Some of these events, such as a guitar maker being subject to an armed Gestapo-like raid by commandos in the employ of the federal Fish and Wildlife Service(!), would make hilarious material for a surreal novel or film satirizing insane bureaucracy. Unfortunately, however, these cases are real, not fiction, and the reality they exemplify is dangerous in the extreme.

According to a 2011 *Wall Street Journal* article that detailed many such exemplary abuses, the roster of federal crimes, which stood at twenty in the original criminal act of 1790, has now grown to at least 4,500 statutory offenses, with unknown thousands of additional grounds for criminal prosecution contained within obscure regulations generated in ever burgeoning amounts by government agencies.

It has thus become impossible for an ordinary citizen to know what is legal and what is not. In fact, as anyone who has ever tried to ensure his or her legal safety by asking for guidance from the IRS or EPA knows, the agencies themselves don't have a clue, and are prompt to disclaim any immunity to prosecution for actions based upon their own advice.

This is an unacceptable situation. A government of secret laws is inimical to both prosperity and liberty. It is very difficult to launch a new enterprise if one cannot be sure that it will be allowed to operate legally. I once held a leading position in a start-up energy company that made equipment to capture natural gas emissions from oil wells so they could be brought to market and sold instead of being flared on-site. It cost us $15,000 just to get a legal determination as to whether US regulations would allow us to obtain a permit to export our gear to Kazakhstan. Not to get the permit, mind you, but just to assess whether such a permit could be obtainable. That's how convoluted the regulations are. Such systems protect monopolies but stifle economic growth, since while major corporations can readily absorb such fees, new small businesses cannot.

Furthermore, freedom itself is at risk since no private person can be safe and secure when a situation prevails in which anyone may face jail for entirely innocent conduct. The Gibson guitar executives were arrested at gunpoint for allegedly using an illegal type of wood in their instruments. How about you? Is your furniture composed of legally harvested wood, or illegally harvested wood?

Unfortunately, in recent years, many judges have chosen to disregard mens rea as an essential principle of basic justice. Instead, to the great convenience of the growing bureaucratic power, they have elected to declare that ignorance of the law—no matter how unknowable—is no excuse, and, trampling on common law, proceed to commit one legal atrocity after another.

What is needed is a forceful reassertion of the rights of free men as defined by English common law. At the core of this is the principle that, since the government exists to protect the life, liberty, and property of innocent people, and not to punish them, it has no business prosecuting them. Thus, mens rea must be established, not as a mere guideline, but as part of the supreme law of the land. No criminal prosecutions should be allowed for offenses that are not widely understood to be crimes.

Such a simplified legal system would not just be essential for Mars. It would be an enormous boon for Earth as well.

Similarly, in civil and contract disputes, the predatory legal caste that currently dominates society on Earth needs to be dethroned or, better yet, eliminated on Mars. This could readily be done by passing a law requiring that all such disputes be settled by binding arbitration. One could go even further and eliminate the arbitrator by having the decision rendered by a jury of three people chosen at random.

Politics is said to be the art of the possible. Reforms along these lines are vitally necessary, but they are so contrary to currently accepted practice that they seem totally unrealistic. Like the ideas of eighteenth-century liberalism that needed to be proven in America before they could be accepted in Europe and Asia, an example is required.

A Martian noble experiment could provide such an example.

Crime and Punishment

Even with a simplified legal system, crime would represent a major problem on Mars. Consider the damage that criminals could do if they were to vandalize or steal critical components of the city's life support system. Putting aside such extreme cases, the smooth functioning of society necessary for its prosperity requires trust between individuals. Teamwork requires a general belief that everyone is on the same side or at least will play by the same rules. Such essential understanding would be fatally undermined by the widespread presence or acceptance of crime.

Furthermore, the cost of imprisoning significant numbers of criminals would be disabling to a Mars city, both in terms of expenditures and the loss of vital talent involved in incarcerating some people and diverting the time of others to guarding and feeding them.

In short, the success, or even survival, of a Mars city would depend critically on its ability to prevent crime.

There are two ways to deter crime: internal or external. That is, people may choose to refrain from engaging in criminal activity either because they think it is wrong, or because they fear the consequences of being caught. The first method, that of conscience, is far more effective, because its enforcer is always present at the scene of the prospective crime. In contrast, police and private security guards are frequently absent, and a cold calculation of potential gain against risk might well indicate theft, assault, burglary, or even murder to be the "smart" move for someone to make given an opportunity.

Moral fabric is mission critical. The strength of the ethical system of a Mars city will be as important for its survival as the soundness of its engineering.

On what foundation can the necessary ethical system be built?

On Earth, religion has frequently served as society's ethical instructor. This can work, as exemplified by the sturdy moral conduct practiced by many people who derive their principles and strength from their faith.

But religion can also be a problem. Religions divide people, and in many cases their practitioners see their moral injunctions as applying only to co-believers, with outsiders viewed as fair game for predation.[8] When viewed this way, religions can act as gang identifiers, reinforcing ethnic in-group bonds. These links establish strong but sharply limited circles of trust separate from that of society at large, and thus frequently serve as the basis for organized crime networks. (This is why various mafias generally have ethnic identifiers, such as the Italian, Irish, Mexican, Roma, Jewish, Russian, Black, Chinese, etc.)

This problem could be surmounted if the entire city were drawn from a single ethno-religious group, but any such city would have limited growth potential and be difficult for any other city to trust.

What is needed is an inspiring, universal, ethical system that every-one can not only accept, but believe in passionately, and aspire to exemplify in their daily lives as a matter of their own sense of purpose and personal integrity. Call it a kind of moral patriotism, if you will. Such a faith is badly needed on Earth today. It will be essential for humans to succeed on Mars.

I am not prepared to prescribe exactly how such a sense of duty can be instilled in a population. I do believe that it is coherent with the combination of individual self-reliance, responsibility, and team solidarity that is necessary for the success of a Mars city in all other areas as well. But there is a tried-and-true method by which it can be found: noble experiments.

Mars cities will take many approaches, consciously or unconsciously, to solve this vital problem. The ones that do the best job will prevail and be copied. The ones that fail to do likewise will disappear.

But while internal checks stemming from a strong civic culture can do a lot to reduce antisocial behavior, they cannot be expected to eliminate crime entirely. What other methods are available?

Well, the most obvious way to reduce the number of criminals is to cut the number of crimes. Human desires will incline people to certain actions, whether they are legal or not. Some of these actions may be morally positive in their intrinsic nature, while others might generally be regarded as negative. Governments have been known to criminalize both. As an example of the former, under the one-child policy implemented in Communist China, parents who chose to have a second child were transformed into criminals, as were their "illegal children," and both were subject to punishment by the state. Examples of the latter include enforcement of laws against the sale and consumption of alcohol during the Prohibition era in the United States, or the treatment of women who engage in marital infidelity in many conservative Islamic countries today.

Martian cities would be very unlikely to wish couples to limit their procreation to one child, but some might well frown upon use of recreational drugs or sex outside of marriage. But given human nature, such behaviors are going to happen. Criminalizing them would create an enormous class of criminals whose keeping would impose a fatal overhead on any city that attempted such a foolhardy experiment. To the extent that such sins need to be deterred, it would be best to let social stigma and the economic penalties that may go along with it do the job and keep the state out of the matter.

That said, there are some actions that are not merely sins, but truly are crimes that need to be suppressed, lest the liberty of all be lost. Take theft or assault, as examples. If anyone can use force to take what is yours or bend you to their will, then you don't really own anything, not even your personal liberty. Humans are descended from animals who engaged in predatory behavior, and there always will be some who, if not otherwise deterred, would be willing to let their baser instincts override their conscience and seek to plunder others. Consequently, organized society cannot exist without some system for punishing criminals.

On Earth, such deterrents have included fines, loss of privileges, public humiliation, corporal punishments, various degrees of enslavement ranging from community service to chain gangs, excommunication, imprisonment, conscription, exile, mutilation, and execution.

Fines and loss of privileges will serve well enough as punishments for mild violations. But for more serious offenses, stiffer deterrents will be required.

For reasons already discussed, imprisonment would not be a practical option on Mars, and humiliation, excommunication, and mutilation would be bad ideas as they would create enraged enemies of society, who, left at liberty, might well seek revenge by destroying everyone else. Exile to Earth would be very expensive. The city might, however, have outposts, such as mining settlements, that might have a need for such undesirables and were sufficiently undesirable themselves as to make being sent there a useful punishment. Exile to the asteroid belt might also work as well.

But how to deal with people who are too violent to be left at liberty? Execution is the obvious answer, but it would be seen as enormously wasteful by societies as labor-poor as Mars city-states are likely to be. That leaves enslavement.

Garnishment of wages, community service, and conscription may all be regarded as degrees of enslavement, yet under certain circumstances could have their place in the legal system of a free society.

But chattel slavery, in which not merely someone's labor, but his or her person itself is expropriated is something else entirely. Some Martian cities may be tempted to adopt it, because selling its victims as slaves is an effective way for a legal system to turn a profit. That is why such practices are embraced by unfree nations on Earth. But they are inimical to the kind of free society that Mars city-states will need to be.

No nation can long endure as half slave and half free. It must become all one or the other. This was the critical question for the young United States and may prove to be so on Mars as well. We shall return to it in our final chapter.

The most successful Martian cities will be the ones that understand that the creative talents of all, including even those found guilty of serious crimes, need to be treasured. They are not the only ones who would benefit from such understanding. There will be no place for prisons on Mars. There really should be no place on Earth for them either. Keeping people in cages where they are trained and brutalized to become even worse types of criminals and organized into criminal networks doesn't make sense. But we continue this insane practice on Earth out of force of custom. Hopefully those starting with a blank slate on Mars will find a better way.

Treatment of the Elderly

We wall criminals out of society on Earth today and feel justified doing so because we regard them with contempt. But how differently do we treat the elderly, whom we affect to respect? Indeed, one of the main scandals of our culture today is the disposal of old people in useless retirement, old-age homes, or elsewhere where they can fade away unseen.

There is a coarseness in our culture today, and it all stems from the same source: We don't value people. We don't value the talents of children, so we shut them away in disciplinary camps we call schools. We don't value the talents of legal offenders, so we shut them away in jails. We don't value the talents of immigrants, so we stop them at the border. We don't value the talent, experience, and wisdom of the elderly, so we shut them away wherever we can.

Mars will need the talents of every person. Successful Mars cities will feature families with plentiful kids. That means most old folks will be grandparents. There are no better teachers. Successful Mars cities will have a patriotic citizenry. That means people who don't take their heritage for granted, who appreciate the grit and guts of those who came before them to build a city and create a nation in a place where previously there was nothing but rocks and dust.

Civilizations become decadent when they lose their grasp of their generative principles, the first and most important of which is the ethos that created them. That ethos is the endowment offered by the old. We discard it at our peril.

The Martians will show us how to keep it.

9

LIBERTY ON MARS

We believe that free labor, that free thought, have enslaved the forces of nature, and made them work for man. We make the old attraction of gravitation work for us; we make the lightning do our errands, we make steam hammer and fashion what we need. . . . Free labor has invented all the machines.

—ROBERT G. INGERSOLL, *Indianapolis Speech*, 1876[1]

ULTIMATELY, I BELIEVE the case for Mars is liberty. It's a very broad thesis, for sure, but in this chapter, I'll try to back it up.

The word "liberty" has originally a Roman root, but it entered the English language through Norman French. There are two words for everything in English—the Norman word and the Saxon word. The Normans were the upper class after the conquest, so the Norman word is the polite word, while the Saxon word is the direct word. So, for instance, you can acquire, or you can take. You can dine, or you can eat. People could be deplorable, or they can be bad. You can possess liberty, or you can be free. Freedom is the direct word for liberty.

When politicians want to appear polished, they employ the word "liberty." But when they want people to really get what they mean, they use freedom. For example, during World War II, the Roosevelt administration wanted to define for Americans what was really worth fighting and even perhaps dying for. They wanted to make clear the cause was not just to get back at the Japanese for their Pearl Harbor attack, but that there were more fundamental principles at stake. So, they set out such principles and enunciated them as "the four freedoms," a piece of political science that everyday people, both in and out of uniform, could truly understand. These four freedoms were freedom from want, freedom from fear, freedom of speech, and freedom of worship. This formulation was quite effective. I still have the wartime letters sent to my mother by my father and her brothers, all of whom were in the army, and they occasionally talk about the four freedoms.

Now, the first two of the four freedoms, freedom from want and freedom from fear, are easy to understand. These are aspirations that are not even unique to humans—we share them with the other higher animals. They are certainly something that any border collie could comprehend. But as smart as they are, border collies might have difficulty with the next two. "Freedom of speech" they might get in a narrow sense, but not in the broader sense of which it's meant, which is political freedom. This is unique to humans, on Earth anyway. It's something that we treasure, and it's something that people today, at least in the West, can still fully understand as something worth fighting and risking life for. The last, freedom of worship, is something of whose importance we have lost deep understanding, because most people in our society—or in any case those who read books of this kind—are secular. So, it needs to be unpacked a bit to make clear why it was seen to be so critical that it was listed along with freedom from want, freedom from fear, and political freedom. I believe this was the case because to the people at that time this was really freedom to actualize your soul, freedom to be who you really are. If it's understood that way, as a specific form of a broader human aspiration—a freedom to realize your life's true purpose in the universe—then I think it's something we can grasp. I believe that human expansion into space is fundamental to the future of all four of those freedoms.

In 1876, Colonel Robert Ingersoll gave a remarkable speech to an audience of Civil War veterans. In my opinion, it is one of the greatest speeches in American history. That speech can still be found in many collections of great rhetoric, sometimes titled the "Indianapolis Speech" or "The Vision of War."[1] In it, Ingersoll tried to explain to veterans of the Union Army what they had fought for. They had defeated slavery. So, he defined the war in terms of the fight against slavery and described the cruelty of slavery. But why were we able to get rid of slavery? How did we defeat slavery? Why don't we need slavery anymore? He says, *we don't need human slavery anymore because the forces of nature are our slaves.* We make the lightning run our messages for us, and we make the steam hammer and fashion what we need. The forces of nature are our slaves, he exclaims. "They have no backs to be whipped, no hearts to be broken." But how have we enslaved these forces, he asks. He answers, through our machines! But where did the machines come from? Where have all these marvelous machines that adorn our age, the steam engine, the railroad, the steamboat, the telegraph, and now in 1876, the telephone, come from? He answers: *from free thought.* Free thought, he says, has given us all the machines! This is the root of all our progress, both material and moral. It has given us wealth, emancipation, and dignity by allowing us to take the stain of slavery off our flag.

Ingersoll's insight is the key to understanding the entire enterprise of Mars. It is free thought that gives us the machines. This is as true today as it was in 1876. The space age began as a contest between the free world and the unfree world, the communists. The race to the Moon would be the test of the two. This is a new ocean, John F. Kennedy said, and we are going to show that free men can sail it—and in fact, they can do it best. Then, despite being behind initially, the United States caught up to the Soviet Union, then surpassed it, and was the first and, to date, the only country to land people on the Moon. The purpose of Apollo was not scientific. It was to astonish the world with what free people can do. I can tell you for a fact that it did just that, because when we landed on the Moon in July 1969, I was in Leningrad playing chess. All the Russians I knew were just amazed and thrilled. *How did you do this? How is this possible?* It made a point. This is something that free people can do.

The system that was set up to enable Apollo certainly exploited the creative potential of a free country to solve all the manifold engineering challenges required to send people to the Moon in the 1960s. But it was imperfect, and it reached its limits. It's true that several unfortunate political decisions in the early 1970s could have been made differently. If they had, we might have proceeded further to establish a Moon base, or even traveled to Mars by the 1980s, as NASA planned to do at the time of the Apollo landings. But ultimately, as a top-down enterprise, it was living on borrowed time. As soon as the political system failed to live up to its responsibilities, it would all have come to a halt. This happened sooner than I would have preferred, but if it hadn't happened under Richard Nixon, then it would have happened in the next administration or the one after that.

So, relatively speaking, after Apollo, our space program stagnated. There was some progress, to be sure. In the following period, a number of important planetary probes were flown. But the stagnation in human spaceflight was remarkable until a new force was activated. This was the potential of entrepreneurial creativity in enabling space activities. We are seeing this unfold now. The price of launch to orbit was stagnant from 1970 to 2010, at $10,000 per kilogram. For four decades this held, like a law of physics that was never going to change. But since 2010, the SpaceX company, led by Elon Musk,—*and motivated by the quest to settle Mars*—has cut the cost of space launch by a factor of five. They did this by first establishing a more rational business structure, eliminating the overhead caused by cost-plus contracting. Then they started introducing partially reusable launch vehicles. They began with the *Falcon 9*, which reuses nine of its ten engines, then followed with the Falcon Heavy, which reuses twenty-seven of its twenty-eight engines. As a result, they cut the cost of a space launch from $10,000 per kilogram to $2,000 per kilogram, and if they are successful in creating the Starship, a fully reusable heavy lift launch system, they will slash the cost of a space launch to less than $500 per kilogram. This is a remarkable thing. But the point here is not only that SpaceX has introduced some very useful, cost-effective launch systems, as well as Dragon capsules and so forth; they've proven a point: It is possible for a well-led entrepreneurial team to do things that, previously, it was thought only the governments of superpowers could do, and furthermore, do it in one-third the time and at one-tenth the cost or less than was previously deemed necessary. Moreover, they also showed that such teams could do things that had been considered entirely impossible—for example, reusable launch vehicles that come back and land on the pad instead of crashing into the ocean. This was deemed more or less impossible by the space establishment, but now it's happened.

SpaceX has set off an entrepreneurial race in the space launch area, in the spacecraft area, and in the spaceflight systems area, all of which are going to serve to bring the cost of these systems down and increase their effectiveness. The cheaper space launches become, the more space launches there are going to be. That means satellites will get cheaper through mass production. The lower the launch cost, the less conservative spacecraft designers need to be. This will accelerate the progress of spacecraft technology. Taken together, this means that spacecraft will rapidly become much more cost-effective.

Furthermore, there's a competition going on that will accelerate these trends. There's some free world competition from companies like Rocket Lab, Relativity Space, and Blue Origin. There's also Chinese competition. There are at least five Chinese entrepreneurial launch companies that I'm aware of that have gotten investor funding and are currently trying to create systems that look a lot like the

Falcon 9. The laws of physics are true for everyone. So, it's a sure thing that they'll be able to do it. Consequently, Musk is going to have to do something better, and, in fact, is trying to do that right now. If Starship is successful, they'll copy that, too, and he'll have to do something even better.

But if the Chinese or Russians ever really want to compete in space, they're going to need to invoke the forces of liberty. The reason there is not a SpaceX in Russia right now is because Russia does not have liberty under law. There is certainly technical talent there, and there are people with large amounts of capital who believe in the importance of human expansion into space. But nobody's willing to start a SpaceX in Russia, because it could be expropriated instantly by the Kremlin kleptocrats. Reason is a stick. If Russia wants to be truly competitive in this arena, they're going to have to create increased degrees of liberty. This will be true all around the world.

Liberty on Earth is something of great importance, but will there be liberty *in* space?

Liberty will be necessary for us to settle space. We will need to create ever cheaper and more cost-effective launch systems, spacecraft, and space transportation systems, and these require liberty. But why will liberty be necessary once we settle space, for example, on Mars?

Some have speculated that there might not be liberty in extraterrestrial colonies because of the degree of individual dependence on governments.[2] The authorities in an extraterrestrial colony could turn your air off and kill you. Therefore, it is argued, these will be totalitarian in nature.

I believe that this thesis is false, for several reasons. First, if you look at the situation on Earth, the populations that are easiest to subject to tyranny are not highly urbanized, highly organized societies. They are peasant societies in which no one individual is particularly essential. That is, the peasant populations and similar rural, nominally self-sufficient populations are the easiest ones for tyrants to tyrannize. In contrast, as the medieval saying goes, "city air makes a man free." In a highly organized society, individual people are more important qua individuals, they're more necessary, and they can do more damage. So, they must be respected more.

There's no one who will need to be respected more than the inhabitant of a Mars colony because he or she could wreck the whole place if they were upset with the government. Dependence goes two ways. Individual dignity cannot be ignored in a highly organized society in which individual human skills and goodwill are at a premium.

But then there's another point, which is, *why would people move to Mars if it meant moving to a place with less freedom?* They would not. There will almost certainly be many Mars colonies founded with different ideas regarding how they should be organized. But the ones that will grow will be the ones that are most attractive to settlers, the ones that offer them the greatest opportunity to realize their human potential. That is why the thirteen colonies and then the growing United States attracted so many more immigrants than Central or South America. They outgrew Latin America because they offered more liberty. There's nothing more valuable to people than liberty. People will cross oceans of saltwater and oceans of space, take enormous risks, and leave their whole lives behind to obtain it. They will not do so to abandon it. So, from a Darwinian point of view, the prevalence of extraterrestrial tyranny is an impossibility because such societies would not be able to grow. They would be outcompeted for immigrants by societies that offer greater liberty.

Furthermore, to grow not just in population, but in prosperity, extraterrestrial settlements will have to offer liberty. Why? Because these extraterrestrial worlds will be frontier worlds. They will be places with a severe labor shortage, which means they're going to need immigrants. That's fundamental, but they're also going to need technology that is unique to their challenges. In the colonial era and the nineteenth century, Americans became famous as gadgeteers, especially in the field of labor-saving machinery, because, once again, an open frontier creates a severe labor shortage. We attempted to deal with that in two different ways. One was through slavery and the other was through invention. Invention won, hands down.

The frontier drove the creation of labor-saving machinery, but also social innovations, such as female schoolteachers, which then propagated East during the labor shortage induced by the Civil War. The frontier labor shortage that induced America's backyard inventor culture is going to be magnified a thousandfold on Mars. The Martians will be innovators not just in labor-saving machinery, but also in their more modern equivalents, which include automation, robotics, and artificial intelligence. Automation and robotics provide means to multiply the quantity of labor available in a population, while artificial intelligence is a way to multiply its diversity of skills. Another critical need on Mars will be energy. Martians will not have access to huge reserves of fossil fuels manufactured for them by billions of years of prior global photosynthetic activity. Only advanced fission and fusion reactors will be able to meet their needs.

The Martians are going to have to be inventors to create these technologies, which will be essential for their survival, and once again, invention requires freedom. As Ingersoll said, free thought gave us all the machines. Furthermore, these inventions will not only meet the local needs of the colonists but will provide patents that can be licensed on Earth and create cash income to pay for necessary imports.

This is the driving force for extraterrestrial liberty. It is necessary for successful settlements. It's essential to create space launch and space transportation systems in the first place to get there, and it's essential to their growth. So, you might say from the point of view of natural selection, only free societies can become spacefaring civilizations. But one can go beyond this.

What *new* kinds of freedom can we gain by going to Mars? We will have the freedom to go to a whole world where the rules haven't been written yet. That will provide an extraordinary opportunity. Writing at the time of the American Revolution, Thomas Paine said, "We hold it in our power to begin the world anew." Extraterrestrial colonists will be able to say the same thing. They will be able to write new rules, to try numerous noble experiments in many places on Mars, and then the asteroid belt.

The British invented the steam engine, but it was Americans who invented the steamboat, because the only highways early America possessed were its rivers, and to navigate them properly, neither oars nor sail power would suffice. For the same reason, Martians will not only aggressively push the development of fusion power, but they will also use it to enable fusion rockets, and fusion rockets will give us interstellar capabilities.

The freedom unleashed on Mars will open the path to the stars.

Noble Hypotheses

The prevailing political culture on Mars will be independent city-states. This will be so not merely because Martian colonists will want to be independent of Earth, but because those cities that are free to reinvest their profits in themselves will perforce outgrow all those whose income remains devoted to pay off Earthside investors. As independent entities, they will be free to experiment with new political forms.

So, what might some of the new world's noble experiments be?

In 1992, philosopher Francis Fukuyama wrote a book entitled *The End of History and the Last Man*, in which he postulated that, with the victory of the Western powers over the Soviet bloc, the perfect form of government had prevailed, effectively ending human history. Developments since that time have rather dramatically refuted Fukuyama's contention that the democratic capitalist world order dominant in 1992 was fated to endure henceforth, unchallenged for all eternity. But there was another aspect of Fukuyama's thesis that was even more questionable, to wit his assumption that the contemporary Western forms of government represented a perfection whose improvement was not possible, or even conceivable. In this respect Fukuyama earned unenviable comparisons to the early-nineteenth-century German philosopher George Frederick Hegel, who in his own work had argued that history had ended with the development of the Prussian bureaucratic state, whose perfect rationality bore no possibility of improvement either.

Hegel's ideal bureaucracy got its marching orders from the Prussian monarchy, whose commands it would implement with mechanical efficiency and integrity, freed through its rigid procedures from influence by either corruption or human sympathy. Fukuyama's allegedly democratic governments are really bureaucracies, too, albeit ruling with some general guidance from elected representatives. In their current incarnations in the more advanced Western countries, these bureaucracies are relatively honest and competent. It is rare, for example, that one hears about the officials at a fire department grafting the money that was supposed to pay for fire engines, or demanding bribes before they will dispatch their forces to put out a fire. Contracts to build government facilities might sometimes go to insiders, however, and payback for favorable regulatory rulings is hardly unknown. The bureaucracy is of mixed competence. There are many sharp and passionately dedicated people working within it. Yet it also includes many deadbeats who really don't care about the job they are supposed to be doing or the people they should be trying to serve. Consequently, as an institution it is frequently stupid or arrogant, and sometimes takes years to respond to urgent public needs.

Then there is the predatory legal caste, whose services are required by anyone who needs to deal with the web of regulations spun by the bureaucracy. Those services are costly.

As the saying goes, freedom isn't free.

For Fukuyama this is as good as it can get. Many Martian colonists will disagree. By going to a new world, they will have a chance to do better. What might they try?

The most frequently suggested alternative to our current combination of representative government, enlightened bureaucracy, and a pay-to-play justice system is direct democracy.

Ancient Athens had a population of three hundred thousand people, and it managed to govern itself via direct democracy. Surely a Martian city-state, which is likely to be smaller, and which would have all the advantages of modern instant electronic communications to enhance public participation, could do so as well.

There is a lot to be said for this. No one knows better than the people themselves what the people want. Bureaucracies, on the other hand, can, and frequently do, ignore urgent public needs, or even enforce policies at direct odds with them, as they act in accord with their own internal fashions and ideologies. This leads to defective governance. But even worse, subjecting the public to rule by such paternalistic overlords degrades the people from free citizens into subjects. This insults and harms the people greatly. It also damages the city itself, because such degradation also diminishes the inclination and ability of its people to deliver the responsibility, initiative, patriotism, and courage a society needs to grow, prosper, or even survive when times get tough.

Yet it must also be said that, as a method of governance, pure democracy has a low batting average. Ancient Athens was a glorious experiment, but it was also a failure, precisely because of the weaknesses inherent in direct rule by the unrestricted will of the people. As the great Athenian historian Thucydides documented in detail is his classic account of the Peloponnesian War[3], the susceptibility of the populace to manipulation by demagogues led Athens to repeated, and ultimately total, disaster. Even worse outcomes can be seen in the French Revolution's Reign of Terror, enforced by Robespierre with the enthusiastic backing of the Parisian mob.

The Founders of the United States were well read in classical history, and so, wishing to avoid the catastrophic potential that unrestricted pure democracy had demonstrated in ancient Athens, proposed a mixed government, dividing power between democratic (the House), aristocratic (the Senate), and monarchical (the Presidency) elements, as well as between the above branches and the judiciary, the federal government and the states, government and individuals, judges and juries, and the present and the past (represented by the Constitution). A Mars city-state may well wish to consider comparable elements.

A monarchical element is arguably necessary to deal with emergencies where fast action is required. But it can easily devolve into dictatorship unless watched over by a body sophisticated enough to know the political game. That is one reason why an aristocracy may be needed. Another is that there needs to be a group of knowledgeable people in the government empowered to take the lead on complex issues. On Earth, this has frequently included foreign policy—which is why America's Founders assigned approval of treaties to the Senate. On Mars this could include dealing with technical matters whose correct handling is vital to the survival of the city. A third reason for an aristocracy is that there simply needs to be a different group of people assessing decisions from the population at large to provide the check of a second opinion against imprudent action embraced by the majority under the impulse of passion.

The idea of an aristocracy is offensive to democratic sentiments, and so it may be rejected by some Martian city-states. Those that embrace it, however, could do so based on any number of criteria.

As an American, blood aristocracy does not appeal to me. But it must be said that it has proved useful to the British, providing their nation with a leadership caste that has seen it through many a crisis.

The Founders' answer to this was to try to create an elected leadership caste, in the form of senators whose long terms of service (and, initially, an indirect mode of election) would provide relief from short-term popular pressure, as well as sufficient time in office to become expert in one or more areas of governance.

But neither a family tradition of service nor public election guarantees excellence in individuals. So, perhaps a city might decide that its Senate should be chosen precisely based on demonstrated contributions. Acting in accord with such a concept, it might create a council of ladies and knights of Mars, women and men of proven worth, whose wisdom, dedication, and integrity could be counted on to provide the city with the best guidance possible.

Another concept that has found its way into several Mars city-state designs[4, 5] would be to create multiple senates in the form of guilds. Drawn from the relevant technical, industrial, or service areas, each of these would have veto powers only over proposals in their own jurisdiction. Thus, everyone would have two voices in the governance of the city: one as a member of the popular assembly and the other as part of his or her guild.

Or a city could just shelve the idea that any group of people, no matter how elected or selected, is more likely to be able to govern better than the rest and just select its ruling council randomly, as juries are currently chosen in the United States. This would offer the advantage of enabling representative government while eliminating elections, electioneering, and both demagogic and oligarchic professional politicians.

Then there is the question of the franchise. Current thinking on Earth is that one person, one vote, with the age qualification set at about eighteen years of age, is the right way to go. But does this necessarily lead to the best governance? America's Founders set a property qualification for voting because they felt that power should be assigned only to those who had a sufficient stake in the community. Modern corporations take this idea much further, by not only restricting votes to shareholders, but assigning the number of votes to each shareholder in proportion to the number of shares. I don't believe that this latter formulation is proper for the governance of a city because, regardless of the size of their share in the city's profits, each citizen has a life of equal worth, and all are impacted by the soundness of their government's decisions. Some people, however, have more than their own lives at stake; they have their children's as well. Perhaps parents should get multiple votes in proportion to the number of children they are raising. Following the Founders' voters-should-have-skin-in-the-game logic, a city might well consider limiting its franchise only to those who are raising, or have raised, children. The justice of such a proposal is debatable, but, as a practical matter, the cities that raise the most children will outgrow the others. So, we might well see such systems in those that emerge as leading metropolises on the Red Planet.

Much less conventional ideas might emerge. In the early twentieth century there was a considerable anarchist movement whose adherents believed that the ideal society would have no government at all. I don't understand how such a system would work. It certainly couldn't on Earth, because no government implies no military, which would leave any such society on our planet helpless against annexation. But positioned on a vast world with no neighboring states, maybe it could. If not the entirety of government, at least all those features that directly or indirectly derive from the military necessity underlying terrestrial governments might be dispensed with. This could lead to very interesting variants, with utopian-on-Earth ideas ranging from anarcho-socialism to anarcho-libertarianism becoming real possibilities.

The Rights of Mars

Any form of government, including most especially direct democracies, can become tyrannical. To be free, therefore, people need protections against government itself. The Founders of the United States set forth this concept as following from Natural Law, wherein *certain inalienable rights* are deemed to have been decreed by *the Creator*—in other words, from a source of greater authority than any government.

The idea of inherent individual rights is a profoundly antidemocratic concept because it gives not just one king, but *every single citizen the divine right to overrule the will of the majority* on at least certain important matters. Yet, while some object to its phrasing, very few Americans would choose to discard the rights it has awarded them. On the contrary, in most cases, they or their ancestors came here— leaving all they knew and loved behind—in large part precisely for the purpose of seeking its blessings.

Martian cities will compete for immigrants. There is no better draw than freedom.

What rights might Martian cities choose to put on offer?

Let's start with the rights of man embodied in the US Declaration of Independence, Constitution, Bill of Rights, and subsequent constitutional amendments. These include:

1. Freedom of Religion, Assembly, Speech, and of the Press
2. The Right to Bear Arms
3. The Right to Due Process and Trial by Jury
4. The Right to Face One's Accusers
5. The Right to be Free of Arbitrary Arrest, Search, or Long Imprisonment Without Trial
6. The Right to Vote for Representative Government
7. The Right to Own Property

8. The Right to be Free of Chattel Slavery
9. The Right to Equal Protection Under the Law regardless of Race, Creed, Color, or Country of National Origin

In addition, there is an emerging consensus and body of law tending toward the establishment of an additional right, which can be phrased:

10. The Right to Equal Opportunity Regardless of Race or Sex

This is the best that twenty-first-century Earth has to offer. Mars cities will need to include it but move beyond it. I think that Martian cities might also choose to incorporate some of the following as fundamental human rights:

11. The Right to Self-Government by Direct Voting
12. The Right to Access to Means of Mass Communication
13. The Right to All Scientific Knowledge
14. The Right to Knowledge of All Government Activities
15. The Right to be Free of Involuntary Military Service
16. The Right to Immigrate or Emigrate
17. The Right to Free Education
18. The Right to Practice Any Profession
19. The Right to Opportunity for Useful Employment
20. The Right to Initiate Enterprises
21. The Right to Invent and Implement New Technologies
22. The Right to Build, Develop Natural Resources, and Improve Nature
23. The Right to Have Children
24. The Right to a Comprehensible Legal System Based upon Justice and Equity
25. The Right to Be Free from Extortionate Lawsuits

26. The Right to Privacy
27. The Right to Free Life Support, Food, Health Care, and Other Basic Necessities

This list will no doubt be controversial, both by what it includes and what it omits. While we don't have enough room here to adequately consider all (or any!) of the inclusions or omissions, let's briefly discuss a few.

Consider rights 11–14 and 24–26, above, whose purpose is to establish an actual democracy—of the people, by the people, for the people. As noted above, this can be a mixed blessing, but provided that the safeguards of divided governmental powers and individual rights 1–10 are also included, democratic governance could be a major step forward. Currently we don't have that. We have a semi-oligarchy with democratic influences. Ordinary citizens have little control over the government, as the bureaucrats appointed by their elected representatives mostly do as they please, and only respond to the public when massive pressure is evidenced. In addition, many government operations are secret, and the legal system is unfathomable. This latter feature exposes the population to looting by a predatory legal caste that possesses exclusive knowledge of how the system works, allowing it to extract tribute from anyone who needs to protect or exercise his or her rights under the law.

Of course, when the United States was founded, indirect representation was the best approximation to democracy that was feasible. But today, with the availability of the internet and other forms of instantaneous electronic communications, there is no fundamental technological reason why the public could not directly engage in voting on legislation, taxation, expenditures, and many other issues. It might be argued that the public is not qualified to do so. Personally, as one who has interacted with some of those calling the shots within the present system, I see no evidence for the public's inferiority. Such skepticism of the people's capacity to participate in direct government is reminiscent of similar skepticism offered by sophisticated European observers of the practicality of the Founding Fathers' notions of the viability of representative democracy, freedom of religion, the press, the right of the people to bear arms, trial by jury, and so on. To the establishment eighteenth-century mind, all these concepts were prescriptions for chaos. It took a "noble experiment" in a new land to prove their viability. Until that was done, it was impossible to implement most of them in Europe. Similarly, the system of bureaucratic governance now prevailing in all advanced Western countries will never yield to actual democracy until the latter is proven somewhere. For that, a new "noble experiment" will be necessary.

Rights numbers 15–23 have all existed, explicitly or implicitly, to one extent or another, at one time or another, in the United States and many other countries. Many of these, however, have become significantly constricted in many nations in recent years, and there will be increasing pressure to do so further on Earth, especially if the increased bureaucratization of government should result in a stagnant zero-growth world. A Martian civilization that offered these as fundamental rights could become the magnet for the dreams and hopes of millions.

Take right number 18, the right to practice any profession, for example. One of most widespread forms of oppression within advanced Western cultures today is denial of the right of the uncertified to exercise their talents. Instead, to enter various professions, our societies demand that individuals obtain certificates from universities or other institutions. As this can only be done at great expense in time and money, such arrangements serve as a kind of society-wide poll tax, maintaining the class structure at a stupendous cost of lost talent.

A Mars city, which needs talent above all, would be very poorly advised to follow our wasteful and reactionary example. Abolishing such restrictions, it would attract myriads of immigrants, and be able to make much fuller use of the talent it gets. It would be a winning move in the race to grow.

But would it be practical? Can a society really function with people free to practice professions without our present set of caste- or guild-like certification systems proving legal distinctions between the rights of the "learned" and the "unlearned"? What about right number 20, the right to implement new technologies, which is also becoming increasing restricted on Earth? Can it be made inalienable? Can a society function with people free to invent and implement new technologies without interference by government regulators, appointed democratically or otherwise?

According to advocates of a new and fashionable Germanic philosophical theory called the "precautionary principle," nothing should be allowed until it is "proven safe." If accepted, this promises to strangle technological innovation on Earth in the name of safety. Can Mars succeed by standing this philosophy on its head? It very well might have to do exactly that.

Perhaps there are other rights that should be added. Take number 27. Could a society function and progress with people guaranteed food, clothing, shelter, medical care, basic income, and/or other necessities as fundamental rights?

I don't know, and I don't think anybody else does either. To find the answers, a lot of noble experiments will have to be run, with various combinations of these or other rights. The ones that work will lead to societies that succeed, and thus be remembered, and copied.

Professor Fukuyama believes that the current administrative bureaucratic systems are the final answer in human social thought. I disagree. I don't know what it is, but I'm sure we can find better. And our best chance of finding it, of taking our next giant step, will occur, when, once again, in a new and as-yet unstructured land, a group of serious-minded people equipped with the experience and most advanced thought of their era gather around a blank sheet of paper, and with all the force of their reason, begin to write:

"We hold these truths to be self-evident . . ."

A Future Without Limits

If we do what we can do in our time, which is establish the first human footholds on Mars to begin humanity's career as a spacefaring civilization, then three hundred years from now, there will not only be a diverse set of new nations on Mars, and among the multitude of the asteroids and other bodies in the solar system, but on thousands of planets, orbiting thousands of stars in this realm of the galaxy. These will be and must be highly innovative, free cultures. That's because only free cultures *can* be highly innovative, and extraterrestrial settlers *must* be highly innovative if they are to be able to deal with the diverse novel environments that they will encounter. So, there will be myriads of new nations with new cultures, new languages, new literatures, fantastical assortments of contributions to social thought, technology, invention, and histories of great deeds that will be used to inspire people to go further. That is something grand and wonderful, *an unlimited future* that we, and all who participate in the great adventure going forward, can be responsible for creating and have the joy of creating.

Freedom will give us space. Space will give us freedom.

That is why the case for Mars is liberty.

10

A GRAND ENGINE
OF INVENTION

There are no rules here—we're trying to accomplish something.

—Thomas Edison

MARS WILL BE A NEW FRONTIER posing novel challenges that can only be addressed by a highly creative and technologically adept culture. To first survive and then prosper, Mars city-states will need to become powerful engines of invention.

This same dynamic interplay between a free society and an open frontier shaped American society, the American character, and America's influence on the world at large throughout its entire history. Its stream started with the first colonial period settlements, became a raging world-changing torrent in the nineteenth century, and has continued to flow as a central factor in global economic life down to the present day. It was key not only to America's success, but to that of the modern global economy that American inventiveness made possible. It will be even more important on Mars.

So, to understand what the future may bring, let's have a look at the past.

. . .

Frontier Ships

We came in ships, so we started with ships.

Beginning as early as the 1600s, English and Dutch colonists in the New World made important innovations in nautical technology by introducing two new revolutionary types of sailing vessels. These were the sloop and the schooner, both with simple fore-and-aft rigs that enabled them to sail an unprecedented forty five degrees into the wind, while meeting the needs of the tight colonial labor market by minimizing the size of the crew required to handle them.

The dominant operating cost of merchant sailing ships was labor. Because they could be run by small crews, schooners survived the faster but much more elaborate clipper ships (also an American invention) to remain an important carrier of maritime trade until World War I swept the last large cargo schooners off the high seas. The sloop rig continues to this day to be by far the most popular rig for small craft.

By the late colonial period, more than a third of the sailing ships built in the British Empire were being constructed in American shipyards. But it was in the application of a totally new force of nature to nautical travel where American inventors really made their mark.

The first steam engine was invented by Thomas Newcomen in 1712 and greatly improved by James Watt in 1765. These devices were static machines, used for pumping water out of the coal mines that were fueling Britain's infant Industrial Revolution. But American inventors, living in a country whose only inland highways were its rivers, quickly saw the potential for a much more exciting application of steam technology.

The first to try was John Fitch, who demonstrated his steam-driven paddleboat *Perseverance* (that's right, just like the Mars rover) to the delegates of the Constitutional Convention meeting in Philadelphia in 1787. Fitch then attempted to introduce his invention into local ferry service. But his steam-driven paddles were suboptimal for propulsion, and he had inadequate financial backing to get through the rough spots, so his business failed.

Robert Fulton's paddle-wheel steamboat design was much better, and his marriage to the niece of Robert Livingston secured him the kind of backing such a revolutionary enterprise really needed. Livingston was a political and financial powerhouse, having served in the Continental Congress on the committee that drew up the Declaration of Independence, swore in George Washington as the nation's first president, and as minister to France, negotiated the Louisiana Purchase on behalf of President Thomas Jefferson. (Most important, he also was the direct ancestor of my darling wife, Hope.) The Livingstons owned vast landholdings in upstate New York and farther west and wanted to see them developed. Moreover, along with many others among America's Founders, Livingston understood the need to radically improve inland transportation to assure the unity of the rapidly growing continental nation. If Fulton's visionary invention could really work, the benefits to both his material and patriotic goals would be enormous.

With Livingston's full-hearted backing, Fulton was able to go big, right from the start. His *Clermont* was 133 feet long and propelled by two Bolton and Watt engines driving two side paddle wheels, set forth on its maiden voyage from New York to Albany on August 17, 1807, carrying more than a hundred paying passengers. To the cheers of astonished crowds lining the Hudson's riverbanks, the *Clermont* steamed directly upstream all the way to its destination, averaging a fantastic 5 mph on the 150-mile-long trip.

Business success was instantaneous, and the company was soon joined by financier Nicholas Roosevelt (grand uncle of later president Theodore Roosevelt), who moved decisively to expand steamboat service from the Hudson to the American West.[1]

The revolutionaries of one era are frequently the reactionaries of the next. Livingston, Fulton, and Roosevelt tried to maintain a monopoly

FIGURE 10.1. *Infant America revolutionized Earth's transportation technology. Young Mars may well do the same for space travel. "Consternation at the Sight of Fulton's Monster," from the 1871 book* Great Fortunes and How They Were Made.[2] *(Credit: Creative Commons)*

on western steamboat traffic, but that proved to be impossible. In 1814, the rival Monongahela and Ohio Steamboat Company launched the *Enterprise*, a superior steamboat. Instead of being driven by low-pressure Bolton and Watt engines like the *Clermont*, she was powered by a novel high-pressure engine devised by American inventor David French. French's engines were simpler, easy to maintain, more fuel efficient, much more compact, and lighter per unit power delivered than the Watt engines. Under the command of Captain Henry Shreve, the *Enterprise* reached New Orleans in time for the decisive battle between American and British forces for control of the lower Mississippi. With her unmatched speed and ability to make headway upriver, the *Enterprise* became Andrew Jackson's favored workhorse, allowing him to rapidly move troops around to counter the British. But when the war ended, Jackson's martial law decree that allowed her to operate in defiance of Livingston's monopoly expired, and Livingston's representatives ordered her confiscated and had Shreve arrested.

Upon getting bailed out, Shreve decided not to stick around for trial. Taking off in the *Enterprise* on May 6, 1815, he steamed directly upriver all the way to Pittsburgh, reaching his destination—2,200 miles *upstream*—in just thirty days. Making headlines everywhere, this epic feat proved the practicality of two-way steamboat transportation across America's vast western river networks. Within a few years, there were hundreds of steamboats operating all over the West, with some steaming up the Missouri River system as far as Montana.

The *Enterprise's* spectacular voyage also proved the practicality of high-pressure steam engines, opening the way shortly thereafter to an even more radical advance in transportation technology—railroads.

With the advent of compact high-pressure steam engines, railroad locomotives became possible. The first ones were built in Britain, but American engineers rapidly took the lead, so that by 1850, there was more railroad track and railroad locomotives operating in the United States than in the rest of the world combined. Together with the steamboat and telegraph, another American invention, they completely transformed the nature of human society. In 1800, as in ages past, most people never traveled more than a day's walk from their home and knew little about the world beyond those confines. A century later, they were part of a global society, linked together by fast transportation, worldwide trade, instant communications, and mass media.

A wildly disproportionate fraction of this innovation was done in America. Nearly all the rest was done in countries, particularly Britain, that were dynamically engaged with America.

The frontier had called into existence a new world.

America continued to innovate forcefully in marine technology throughout the nineteenth century, inventing both transatlantic steamships and the fast clipper ships that for a while raced them for dominance in world trade. In both inventions, the need for speed was paramount. This was important to improve the economics of all kinds of long-distance trade, but especially key to those seeking to bring in large numbers of immigrants, since the faster the ship, the cheaper the required quarters could be. Then, in the twentieth century, we advanced sea travel further by creating oil tankers, Liberty ships, fast container ships, and nuclear-powered ships, and transcended it entirely by inventing a totally new technology that allowed people to speed anywhere on Earth by flying through the ocean of air.

I believe that the Martians will play the same role with respect to space travel.

The shipyards of Mars will probably begin with shops for repairs and spare parts. Such services would clearly command top dollar from any arriving ship that needs them to be able to return to Earth. Those unable to pay would soon become sources of space parts for the rest. If the right replacement parts are unavailable, scrapped ships could provide the high-grade alloys to make them, and if that source proves insufficient, metals, plastics, and other materials of Martian manufacture will be made to fill the breach. Colonists buying used Starships or similar vessels for one-way trips to Mars will not need them after finding permanent living quarters. Instead of being scrapped for spare parts, the best of those that make it to the Red Planet will likely be sold on arrival to be refitted for local use, including long-distance surface-to-surface transportation on Mars and exploratory mining ventures to the asteroid belt.

The main asteroid belt lies between Mars and Jupiter. It contains thousands of small bodies that are far richer in platinum group metals than the best ore on Earth. Mining them for export to the terrestrial market could become a very big business, but as I have shown,[3, 4] the logistics of supporting such operations from Mars would be orders of magnitude easier than doing so from Earth.

Mars will serve as the hub for supporting the commercial exploita-tion of the asteroids in much the same way that the New England colonies did for the highly profitable Caribbean sugar plantation islands during the period of the colonial Triangle Trade. In conse-quence, just as they did in New England, shipyards should blossom on Mars.

The first Martian asteroid mining vessels will probably be refits of ships previously used to transport immigrants or cargoes to Mars. But such designs, while usable for asteroid mining, are not optimal. So, the Martians will start modifying them. At first the modifications will be relatively modest reconfigurations of Starships or related types of Earth–Mars transports, comparable in their significance to the new sailing rigs that colonial Americans invented.

But the propulsion issues associated with asteroid mining are fun-damentally different from those of Earth to Mars transportation. A high-thrust rocket like Starship could travel from Earth to Mars on a Keplerian elliptical transfer orbit in six months with a ΔV kick of 4.2 km/s out of low Earth orbit, then aerocapture at Mars and land with a further ΔV of about 0.5 km/s. In contrast, such a ship leaving Mars orbit for a destination like Ceres in the main asteroid belt would need a ΔV of about 7.7 km/s for the round trip (4.9 km/s for the out-bound, 2.8 km/s for the return), with an eighteen-month transit each way. Furthermore, while the launch window from Earth to Mars using such Keplerian (impulsive boost followed by long unpowered glide) orbits occurs every two years, the window for the Mars–Ceres trip would only open once every three years.

In short, while such high thrust chemical rocket ships employing Keplerian transfer orbits are acceptable for initial Mars-based asteroid mining businesses, in the long run they would be suboptimal. For that purpose, continuous thrust systems are needed.

Nuclear or solar electric propulsion systems employing ion drives could be considered. But sunlight in the asteroid belt is only about one-seventh as strong as it is on Earth, sharply limiting the potential utility of solar electric propulsion for asteroid mining. Space propulsion nuclear reactors, on the other hand, need to be fueled with highly enriched uranium or plutonium, whose manufacture requires an extensive industrial infrastructure unlikely to be available for a long time on Mars. (Molten salt reactors could be used, but they would be heavy.)

The right answer is fusion propulsion. The fuel for thermonuclear fusion is deuterium, or heavy hydrogen, which is five times as plentiful within Martian water as it is in that of Earth. Because it weighs twice as much as ordinary hydrogen, the enrichment of deuterium from ordinary hydrogen is much easier than separating fissile uranium-235 from non-fissile uranium 238 (the isotope that comprises more than 99 percent of natural uranium), which only differ in weight by a little more than 1 percent. Just getting the natural uranium feedstock for the fissile fuel enrichment process requires prospecting and mining, to obtain ore, which then needs to be transported and subjected to chemical processes of considerable complexity before the enrichment can even begin. In contrast, deuterium can be reliably found in every gram of water on Mars, and the life support system of any Martian settlement will electrolyze a great deal of water every day. This will automatically create large amounts of pure hydrogen gas that can be fed to the enrichment system.

Earth offers numerous attractive alternatives to fusion power, including not only nuclear fission, but fossil fuels, hydro power, wind, and solar. Solar energy can be used to drive robotic probes. But sunlight on Mars is only 40 percent as strong as it is on Earth, and Mars's susceptibility to dust storms that could cut off solar power altogether for months on end makes it too unreliable to consider as the primary power source for Martian city-states. The rest of the list of nonnuclear terrestrial options wouldn't work at all.

Nuclear fission will work and will certainly power initial Mars exploration missions and early settlement. But as settlement proceeds, reactors that can be fueled from Martian resources will become increasingly desirable. In the medium term, homogenous molten salt thorium breeder reactors like the lithium fluoride thorium breeder reactor (LIFTR) design pioneered by Alvin Weinberg at Oak Ridge in the 1960s could meet the needs for surface power of the first generation of Mars city-states.

But the best energy technology to support the large-scale settlement of Mars is fusion power. Driven by spirited national competition between the Americans, Soviets, Europeans, and Japanese, the world's principal government-led fusion programs made great strides from the 1960s through the 1990s, increasing the performance parameters of experimental fusion devices by a factor of one hundred thousand over the period in question.[5] Raising performance another factor of four would have allowed the devices to achieve ignition, a condition in which fusion reactions in the magnetically confined ionized fuel gas would maintain its temperature, allowing unlimited energy to be produced with no additional input power.

Unfortunately, this progress was brought to a dead halt by the decision of the bureaucrats leading the four principal competing government programs in the mid-1980s to collapse their efforts into a single joint program, called ITER (for International Thermonuclear Experimental Reactor). With competitive pressure thus removed, government-backed fusion research since then moved forward at a snail's pace—or rather somewhat slower, as four decades later the machine has not been completed, let alone turned on.

However, the success of SpaceX has led some serious investors to suspect—correctly in my view—that, as for the case of achieving reusable space launch, the fundamental problem afflicting the fusion program was not technical, but institutional. Where governments failed,

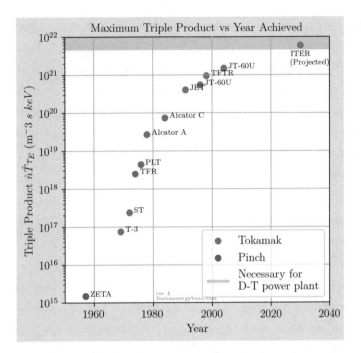

FIGURE 10.2. *Progress in fusion energy. Since 1965, the world's fusion programs have advanced the achieved Lawson Triple Product (density x temperature x confinement time) by a factor of one hundred thousand. A further increase of a factor of four will take us to ignition. This awaits the building of new experimental machines, which halted in the 1990s. Note the logarithmic scale. (Credit: Fusion Energy Base)*

they concluded, entrepreneurs might succeed. As a result, several entrepreneurial fusion efforts have received substantial funding—amounting, in several cases, to hundreds of millions of dollars. These companies are moving ahead fast, and it is likely that one or more of them will achieve ignition this decade.[6]

Consequently, it is likely that first-generation commercial fusion reactions will become operational on Earth not long after serious efforts to settle Mars will begin.

So, early on, Martian city-states will include fusion reactor engineers in their ranks. These will be necessary to run, maintain, and repair the settlement's primary power systems. But, being engineers, they are sure to start tinkering with them to improve performance. Furthermore, since they will be living on a planet that stands to vastly expand its income potential if fusion reactors can be adapted from spaceflight, they will devote passionate efforts to solve that problem. In doing so they—like the nineteenth-century engineers who revolutionized stream engines by redesigning them into compact high-pressure designs needed to drive mobile systems—will radically improve fusion power itself.

As steamboats were for the American frontier, so fusion ships will be for that of Mars. As steamboats were for steam engines, so fusion ships will be for fusion power.

A fusion rocket would differ from the doughnut-shaped tokamaks that are currently the lead technology for first-generation commercial fusion reactions. The high temperature ionized gases, or plasmas within which fusion is performed, are contained by limiting their movement using magnetic fields. Plasmas can move freely along magnetic field lines but have difficulty crossing them. The tokamak exploits this phenomenon by arranging its magnetic field as a toroidal

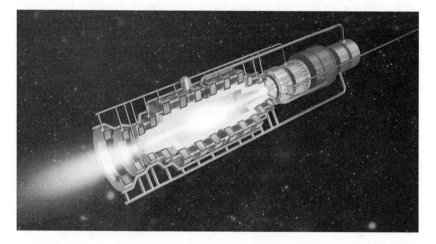

FIGURE 10.3. *A fusion rocket would operate by allowing its super-hot plasma exhaust to escape out of one end of its magnetic confinement chamber, thereby producing thrust. (Credit: NASA and Princeton Satellite Systems)*

racetrack. A fusion rocket, on the other hand, would probably be cylindrical in form, with the magnetic field running down the cylinder's center line. Plasma would be free to run out one end, while escape out the other end would be blocked by magnetic or electrostatic mirrors.[7] The stuff that escapes would constitute the rocket exhaust.

Using pure deuterium plasma, the exhaust velocity produced could be as high as 18,000 km/s, or 6 percent the speed of light. This would create a marginal capability for interstellar travel.[8] But the ΔVs needed for fast round trips from Mars to the asteroids are only on the order of 50 km/s and having an exhaust velocity a few times this would be more than adequate to keep fuel mass to a small fraction of total mission mass. The ship would therefore probably heavily dilute its exhaust with ordinary hydrogen, drastically cutting fuel costs while multiplying its engine's thrust.[9] By diluting its exhaust 10,000 to 1 by mass, for example, it would multiply its thrust a hundredfold and still maintain a spectacular exhaust velocity of 180 km/s.

Such cylindrical reactors would not only be the right way to do fusion propulsion, but they would also be a better design for fusion electric power systems as well. Why? Because their cylindrical shape would be superior to a toroid from an engineering point of view, and because the power of their high speed charged exhaust could be directly converted to electricity at more than double the efficiency (more than 80 percent compared to less than 40 percent) of the steam cycles required by the toroidal system design. In addition, the ultra-high temperate exhaust of cylindrical reactors could potentially be used as a "fusion torch" that could turn any piece of solid material into plasma, whose ionized components could then be separated by magnetic fields.[10] The industrial applications of such technology would be practically unlimited.

Fusion reactors of great size and power could be built in space because they would not face the limiting constraint of providing vacuum vessels. Getting rid of the vacuum vessel would not only save its construction, maintenance, and replacement costs, it would also eliminate the problem of cooling the reactor chamber first wall. Making the reactors big is also key to attaining high temperatures and high efficiency because the confinement time of a magnetic fusion reactor increases in proportion to the square of its characteristic dimension.[11]

Developed first as the uniquely advantageous technology for rocket propulsion, large cylindrical orbiting fusion power reactors could potentially beam their power down to Earth or Mars. Alternatively, such systems (and their component technologies, such as high temperature superconducting magnets) could be first engineered for use in the favorable space environment, which (as steamboats and railroads did for steam and submarines did for nuclear fission) would provide a nursery for their further perfection into more compact forms suitable for operation on the surface of the Earth or Mars.

Aside from LIFTRs and fusion, another energy technology that Martians are likely to pioneer is dry hot rock geothermal energy. Mars is geologically active. Its large shield volcanoes were formed only four hundred million years ago, which is less than 10 percent the age of the planet. If there were sources of geothermal energy capable of producing them, then they almost certainly still exist today. These could provide reliable sources of energy in the multi-megawatt range. That is not enough to power a city of any size but would be very useful to support smaller settlements and isolated outposts and homesteads too small to justify a fusion reactor. By developing such technology to meet their own needs, Martians could provide a valuable advance for use on Earth, as well.

Biotechnology

Land was not in short supply on the American frontier, but labor power to farm it was. In consequence, American frontier agricultural technology focused on labor-saving machinery, with its world-changing innovations including the shearing (or "singing") plough, the McCormick Reaper, threshers, harvesters, and later, tractors. Water was also an issue, which was dealt with by developing new irrigation technology, notably the iconic windmill-driven water wells that adorn America's plains. Frontier agriculture also faced a major problem in getting its harvest to market. This obstacle was remedied by Yankee ingenuity's overall drive to improve transportation technology, including steamboats and railroads in general, and refrigerated transport in particular.

In contrast, Martian frontier farming will need to deal with a shortage of land, water, and labor. Its own city-state market will be close by. But it will need to overcome transport challenges if it is to supply food to support asteroid miners or other distant outposts as well.

Martian agriculture will achieve extreme water conservation by completely recycling the water used in its greenhouse farming system. But the very fact that it is confined to greenhouses will sharply limit the acreage available to Martian farmers. Consequently, there will be a much more powerful incentive to maximize productivity per area than exists anywhere on Earth today.

Terrestrial agriculture has achieved huge gains in productivity over the past two centuries, and particularly since the end of World War II, through the increasingly effective use of irrigation, fertilization, pesticides, and crop breeding. As a result, while the world population has quadrupled over the past eighty years, people everywhere are eating far better. When I was a small boy in the 1950s, my parents would urge me to finish my breakfast because "children are hungry in Europe." Now, not only is hunger in Europe unthinkable, but it is virtually gone from China and India as well. In 2021, the state of Iowa alone produced more corn *than the entire United States did in 1947*—and we were already an agricultural superpower at that time. We have indeed come a long way.

In the logistic sections presented in Chapter 6, I based my estimates for the crop areas required to support a city based on such high efficiency modern forms of traditional agriculture. But greatly increased yields per hectare could be obtained using vertical farms, aeroponics, hydroponics, and aquaponics, provided that cheap large-scale sources of power for indoor illumination and advanced automation systems to reduce associated labor can be developed.

Even more could be accomplished through genetic engineering. While staunchly opposed by self-professed environmentalists, genetically modified crops offer tremendous environmental benefits, allowing more food to be grown with much less land clearance, tillage, irrigation, fertilizer, and pesticides. By making crops more productive, less land is needed to meet humanity's demand for food. By providing plants with precisely selected spliced-in genes, crops could be made more drought-resistant, thereby reducing the need for irrigation, or be given the ability to produce their own nitrates or pesticides, thereby eliminating the need for chemical fertilizers or insect sprays. In other words, essentially all the agricultural issues that the environmental movement has been complaining about for decades could be addressed by the genetic engineering of crop varieties.

Furthermore, far from representing a more dangerous type of food than traditional varieties, genetically modified crops are far safer, precisely because their content can be controlled. Except for fish and game, humans today actually eat no "natural" foods. All our domestic plants and animals are the product of thousands of years of artificial breeding, radically altered from their natural state. That is very fortunate because wild plants and animals are frequently quite dangerous, as it behooves them to be if they are to survive in the wild. It is not genetically modified tomatoes that represent a threat to the consumer, but wild tomatoes, which are loaded with toxins and allergens—so much so that in early-nineteenth-century America, daredevils could sometimes make themselves a hatful of cash by eating a tomato in front of a crowd on a bet. A traditional plant breeder seeking tougher survival traits for agricultural tomatoes might try to crossbreed them with a wild variety, but in doing so he would risk importing genes into the crop that could produce potentially lethal substances at random. In contrast, the genetic engineer can act with precision and select only the traits desired. Not only that, the genetic engineer also has the freedom to import beneficial traits from other species too distantly related to allow combination through conventional breeding techniques.

The benefits following from this ability are enormous. The yield per acre of American corn growers has been radically improved using genetically modified strains. Since the 1940s, the yield of corn bushels per acre has risen nationwide from thirty bushels to an average of almost 180 bushels today. As the best farms are now achieving harvests approaching three hundred bushels per acre, continuation of this progress may allow yet another doubling of yield within the next decade or two. Not only that, but experiments are now underway with corn varieties containing genes that give them the power to fix their own nitrates—an improvement that, once implemented, will radically reduce the amount of chemical fertilizer needed to grow corn worldwide, save energy, and greatly increase yields in Third World countries that are too impoverished to currently use fertilizer. Drought-resistant varieties of cassava and several other staple crops essential to Africa have been developed, which promise to do much to ease the threat of famine to the poorest of the world's poor. Biotechnology can also be used to produce genetically enhanced trees, which grow faster and straighter than unimproved varieties, thereby allowing them to produce much more useful high-quality lumber on a given amount of land. By modifying genes to cause growth hormone to be produced continuously, fish have been genetically engineered to grow much bigger and more quickly than wild strains. This could greatly improve the productivity of aquaculture.

And far from being less nutritious than "natural" (conventionally bred) crops, the new varieties are more nutritious—in some instances, radically more so. To take one important example, by placing beta-carotene-producing genes from daffodils into rice, genetic engineers have created a new strain called "golden rice" rich in vitamin A. Around the world, 250 million children suffer from vitamin A deficiency, with perhaps 500,000 of them going blind every year, and as many as 2.5 million children dying each year because of the deficiency. The planting of golden rice in place of conventional rice could end that horror. Another variety of genetically engineered rice has been developed, and it is rich in iron, offering the prospect of ameliorating the anemia that impairs the lives of more than half a billion women worldwide, mostly in poor countries. Bioengineered crops can also be created to immunize people against diseases.

The implementation of these miraculous innovations is being stalled on Earth by demagogues warning of mythical dangers working in alliance with protectionists seeking to secure backward local agriculture against competition. But the pragmatic Martians will have no time for such reactionaries. They are sure to become virtuoso practitioners of genetic engineering. To multiply the productivity of their greenhouse and hydroponic farms, they will create numerous highly beneficial plant and animal strains, in the process discrediting the green technophobes holding back vital progress on the home planet.

But Martian biotech inventions will not be limited to introducing genetically improved species into relatively conventional, if highly automated, greenhouse, aeroponic, hydroponic, and aquaponic farms. Rather, they can be expected to launch radically new types of microbial bioengineering. For example, using genetically engineered bacteria fed with carbon dioxide and methanol, biotech reactors can be created that produce protein-rich food in vats, offering crop yields from a 50 m^3 vessel (a sphere 4.6 meters in diameter) matching those from twenty-five acres of soybeans,[12] with harvests beginning within a few days of start-up. Producing a substantial fraction of a Mars city's food in this way could greatly reduce the amount of greenhouse land needed to support the system by traditional forms of agriculture. Numerous other products can also be produced by design in such reactors using genetically engineered bacteria, including cellulose, starch, silk, leather, drugs, biofuels, and plastics.

A widespread notion in contemporary science fiction is that in the future humanity will possess "nanotechnology," which will enable self-replicating microscopic robots that can be programmed to perform any task. Bacteria are such self-replicating machines in reality. We are now able to read their genetic code and are learning how to write it. This will allow us to program them, offering the prospect of a new industrial revolution to those willing to grasp its potential.

The availability of cheap land and fossil fuels on Earth has held up development of such programmable microbial technology here, deterring investment by presenting a high threshold of competition from well-established conventional sources. On Mars, traditional agriculture would not be a factor, and these infant but infinitely promising new methods of production would be given room to take wing.

The theoretical efficiency of photosynthesis is 4.5 percent. The actual energetic efficiency of a well-run Iowa farm producing corn from sunlight is about 0.25 percent. Imagine what could be accomplished if plants could be engineered that could approach the theoretical maximum, or if the efficiency of photosynthesis itself could be raised. The gains to humanity on Earth—and to the Martians filing the patents—could be incalculable.

The fusion torch will be able to disassemble all kinds of matter into its component atoms. Programmable biotech will provide the ability to reassemble atoms into any compound desired. By combining these two technologies, Martian civilization will open the way to unlimited material resources for all of humanity, forever.

Unleashing Minds for Labor

Frontier America's labor shortage drove the invention of labor-saving devices—from farm machinery, construction machinery, factory machinery, power tools, and machine tools, to telegraphs, sewing machines, typewriters, and elevators. The inventiveness went further, creating the ultimate open-source platform for labor-saving devices: centrally generated electric power. From this followed, among other things, dishwashers, and washing machines and other home appliances that liberated half of the population from most of the drudge work that was consuming its time and talent. Human materials living standards are defined by the total product available per person. This must on average also be equal to the amount produced per person. By multiplying the productivity of labor, Yankee ingenuity's constant gadgeteering dramatically raised real wages far beyond anything possible in the Old World, making America an ever more compelling destination for immigrants.

Martian city-states will face a labor shortage far more extreme than anything encountered in America. Their drive to multiply the power of labor, both in magnitude and diversity of skills will accordingly be even more forceful. The Martians will therefore not only be expected to excel in applications of labor-saving machinery and automation but go well beyond them to fully exploit the new fields of robotics and artificial intelligence with an enthusiasm that terrestrial society is unlikely to match. Instead of trying to "preserve jobs," Martians will seek every means possible to eliminate them. In short, they will bring on the robot revolution.

Mars has been sometimes described as "the only planet inhabited entirely by robots." However, *Spirit, Opportunity, Curiosity, Perseverance*, and their kind are not really robots. They are telerobots, which is just a fancy term for remote-controlled machines. Similarly, while it is sometimes said the work of the highly automated assembly lines that have become dominant in the automobile industry since the 1980s is being done by robots, this is not true either. Such industrial systems are not robots, but examples of advanced automation using combinations of highly programmable machine tools.

The early-twenty-first century has also seen the proliferation of computer programs, like Siri, Alexa, Google Search, and the voice in your smartphone GPS app, which can advise you how to drive through a complex system of side streets, detours, and highway on- and off-ramps to reach your destination. Their electronic advisors may fairly be described as expert systems. Artificial Intelligence systems like ChatGBT are now coming on the market with impressive capabilities. But they are not robots either.

Martians will certainly exploit telerobots, advanced factory automation, expert systems, and AI to the maximum extent possible. But true robots, combining mechanical systems with artificial intelligence to create a capability approaching that of a human being, are still in an embryonic stage of development. A true self-driving car that can reliably negotiate its way through traffic in all weather, dealing with potholes, squeegee-men, speed-trap cops, deer, and road-rage or alcohol-impaired nutcase drivers, would be a step in that direction. But viewed in an evolutionary context, if we ignore the fact that it can't fuel, fix, or reproduce itself, it would approximate a true robot about as well as a fish approximates a human being. It would have the ability to move around, recognize objects, and respond to them appropriately. But the range of its abilities would be extremely limited. Such systems would certainly be used on Mars to eliminate the need to waste humans on such tasks as local or long-haul delivery drivers. But if such a self-driving truck carrying ore from an outlying mine to the city broke down along the way, you would still need a human crew— or its true robotic equivalent—to go out and fix it.

The idea of robots that can do what a human can do has been a staple of science fiction for more than a century. It is still some distance off, but it is coming. All inventions are combinations of previous inventions. To create a true robot, we would need to combine innovations drawn from many areas, with the three topping the list being electrically actuated locomotion, sensors, and artificial intelligence. All three of these are currently advancing separately by leaps and bounds. Efforts to combine them are inevitable. Boston Dynamics recently demonstrated the ability of its bipedal robot-like machines to breakdance. Impressive as that is, if left at that, such devices are still just fancy puppets. You could use one to put on a tap dance show, but you couldn't send it to the store to buy a bag of carrots. But install an artificial intelligence program into it a bit more sophisticated than that needed for a self-driving car, and you would have something that could do the job of an errand boy. Advance the AI, and its utility for employment would expand exponentially.

What form will robots take? For certain kinds of jobs, nonhuman forms are optimal. If you want a robot for cleaning out pipelines or ventilation ducts, a snakelike form would suit. For undersea work, an otter-like form, combining excellent swimming abilities with good manipulative arms and hands, might well be best. Systems designed to duplicate the roles of field pack animals might well be designed in similar forms to dogs, llamas, horses, or mules. But for the most versatile operation within a human social environment, the human form would be best. It would allow the robot to use all human tools, ranging from wrenches and cooking ware to bicycles and roller skates. It would allow the robot to travel through buildings and in vehicles designed for humans. And it would enable robots to fill in for humans in any task that ever-advancing artificial intelligence will allow.

The GPS in my phone gives me driving directions in a feminine voice with a smart BBC British accent. My wife prefers a manly Australian in hers. While not required to fulfill their functions, similar sales imperatives will probably drive designers to make their robots look and talk like humans as well. People going into a store are likely to be much more comfortable dealing with a charming robotic shopgirl than a monstrous metallic praying mantis. Store owners will buy their staff accordingly.

Such humanoid robots will not need to be programmed to do each new job by teams of nerds writing millions of lines of code. Instead, the combination of sensors and artificial intelligence incorporated into their fundamental design will allow them to be taught much as humans are taught. However, they will learn faster because their perfect memories will allow them to be trained without the tedious repetition of instruction that human rookies generally require to take on a new job.

Most jobs that can be largely carried out using a fixed set of skills will become open to takeover by such robots. This possibility will be widely viewed with considerable trepidation on Earth, and those fears will no doubt greatly delay the introduction of such technology on the home planet. But on labor-short Mars, it will be welcomed with enthusiasm.

Beyond those general outlines, the specific forms that robots, and robot-assisted human culture, will take on Mars must be matters of speculation. But speculation is fun. So, here's my guess.

One prediction might be that, for reasons of mass production economics, robots will be manufactured in a limited variety. Perhaps there will be as few as two general types, corresponding to feminine and masculine forms. Call them Robomaids and Robomechanics, with each optimized to fill the roles most commonly associated with their gender.

That would be logical. But humans are playful, and some robot manufacturers would probably seize upon that aspect of the human psyche to make a bundle offering more variety. Complete one-of-a-kind robots built to order might be too expensive for most customers, but why not at least offer a catalog of off-the-rack choices of robot forms? These could then be uploaded with any specialty program desired. Such forms could be based on well-known historical figures, or pre-2023 movie actors who failed to have their likenesses copyrighted.

So, for example, customers with a twisted sense of humor might want to impress their guests by owning a robotic house servant made to look like Napoleon Bonaparte or Donald Trump. More practically, if you are running a retail store, why settle for a generic salesgirl when, for a little extra, you can have Lauren Bacall? ("If you need anything, just whistle.") Arnold Schwarzenegger, everyone's favorite robot, is sure to be a best-seller for companies offering mine or roadside rescue services. ("Come with me if you want to live.") But rescue units based on Gal Godot's Wonder Woman role might also be quite competitive.

I leave it to the reader to think of additional possibilities.

Whether approached in a utilitarian or playful manner, however, the overall goal would be to free humans from doing any task that robots can do just as well. Robots that do all the housework and assist greatly with childcare would make Martian women the envy of their sisters on Earth, as they will be free to have as many children as they like and still have careers. That would be a strong factor in accelerating the growth of a Martian city, as it would serve the triple imperatives of encouraging female immigration, high birth rates, and the mobilization of feminine talent outside of the home.

Robots will likely ultimately do nearly all the hard and risky work that will need to be done in the potentially lethal Martian outdoor or underground environments. Martian city-states will need to do a lot of mining and tunneling, and these activities are intrinsically dangerous, because geological structures are designed by chance. Transport across the Martian outback, whether to distant destinations on Mars itself, or out to the asteroid belt, will also be intrinsically risky. Historically, such dangerous jobs have generally gone to men, as they are less rick adverse and more expendable, from a reproductive standpoint, than women. This will probably be the case on Mars as well. That said, to the maximum extent possible, the risk to men should be reduced, too. This could be done by having most of the outdoor and underground work done by platoons of robots, with the men serving as their officers, only needing to intervene personally into the dangerous environments when occasion requires it.

So, some men and women will continue to do "jobs," with their productivity greatly multiplied and risks sharply cut by serving as supervisors providing guidance and leadership to the robots doing most of the actual work.

But the highest aspiration of human beings should not be to do "jobs," which is to say to perform tasks dictated by others in exchange for the means of subsistence. It should be to be free to create as they will. The robot revolution pioneered on Mars will make that possible.

11

TRANSFORMING MARS

LIFE CREATES NATURE.

Mars today does not offer the optimal environment for humans or most other forms of terrestrial life. So, human settlers will seek to improve it.

Some people consider the idea of transforming, or "terraforming," other planets to be heretical—humanity playing God. Others would see in such an accomplishment the most profound vindication of the divine nature of the human spirit—dominion over nature, exercised in its highest form to bring dead worlds to life. Personally, I prefer not to consider such issues in their theological form, but if I had to, my sympathies would be with the latter group. Indeed, I would go further. *I would say that failure to terraform constitutes failure to live up to our human nature, and a betrayal of our responsibility as members of the community of life itself.*

These may seem like extreme statements, but they are based in history, about four billion years of history. The chronicle of life on Earth is one of terraforming—that's why our beautiful blue planet is as nice as it is. When the Earth was born, it had no oxygen in its atmosphere, only carbon dioxide and nitrogen, and the land was composed of barren rock. It was fortunate that the Sun was only about 70 percent as bright then as it is now, because if the present-day Sun had shined down on that Earth, the thick layer of carbon dioxide in the atmosphere would have provided enough of a greenhouse effect to turn the planet into a boiling Venus-like hell. Fortunately, however, photosynthetic organisms evolved and transformed the carbon dioxide in Earth's atmosphere into oxygen, in the process completely changing the surface chemistry of the planet. As a result of this activity, not only was a runaway greenhouse effect on Earth avoided, but the evolution of aerobic organisms that use oxygen-based respiration to provide themselves with energetic lifestyles was enabled (though a primeval EPA dedicated to preserving the status quo on the early Earth might have regarded this as a catastrophic act of environmental destruction). This new crowd of critters, known today as animals and plants, then proceeded to modify the Earth still more—colonizing the land, creating soil, and drastically modifying global climate. Life is selfish, so it's not surprising that all the modifications that life has made to the Earth have contributed to enhancing life's prospects, expanding the biosphere, and accelerating its rate of developing new capabilities to improve the Earth as a home for still more life.

Humans are the most recent practitioners of this art. Starting with our earliest civilizations, we used irrigation, crop seeding, weeding, domestication of animals, and protection of our herds to enhance the activity of those parts of the biosphere most efficient in supporting human life. In so doing, we have expanded the biospheric basis for human population, which has expanded our numbers and thereby our power to change nature in our interest in a continued cycle of exponential growth. As a result, we have literally remade the Earth into a place that can support billions of people, a substantial fraction of whom have been sufficiently liberated from the need to toil for daily survival that they can now look out into the night sky for new worlds to conquer.

It is fashionable today to bemoan this transformation as destruction of nature. Indeed, there is a tragic dimension to it. Yet it is nothing more than the continuation and acceleration of the process by which nature was created in the first place.

Life is the creator of nature.

Today, the living biosphere has the potential to expand its reach to encompass a whole new world on Mars, and the interplanetary civilization that develops as a result will have the capability of reaching much further. Humans, with their intelligence and technology, are the unique means that the biosphere has evolved to allow it to blossom across interplanetary and then interstellar space. Countless beings have lived and died to transform the Earth into a place that could give birth to a species with such capabilities. Now it's our turn to do our part.

It's a part that four billion years of evolution has prepared us to play. Humans are the stewards and carriers of terrestrial life, and as we spread out, first to Mars, and then to the nearby stars, we must and shall bring life to many worlds, and many worlds to life.

It would be unnatural for us not to.

How It Can Be Done

Mars was once a warm and wet planet. Evidence of past water action—dry ponds, lakes, streams, and rivers—is to be found everywhere on the Red Planet. Our robotic probes have found salt-encrusted shores of ancient lakes, and detected the basin of what was once a great northern ocean. Many other reminders of the past presence of liquid water, including not only salt deposits but sedimentary and conglomerate rocks, have been found as well.

Yet today, not a drop of liquid water exists anywhere on the Martian surface. The water is still there, of course, oceans of it, frozen in the form of ice caps, glaciers, or permafrost, but except for hydrothermal subsurface reservoirs, all of it is too cold to flow.

Mars's water was liquid in the olden days, because at that time, its carbon dioxide atmosphere was much thicker, and this provided it

234 | THE NEW WORLD ON MARS

with the benefits of a powerful global greenhouse effect. So, the planet was made warm enough for an active water cycle, complete with rivers, lakes, oceans, and rain. But when the water rained down through the carbon dioxide air, it would capture some of the gas in solution, and then react it with soil to form carbonate minerals. This happens on Earth, too, but its vast interior reservoirs of geothermal energy drive our continents to migrate through a process known as "plate tectonics." This constantly causes Earth's crustal material to be subducted underground where the heat breaks down the carbonates, thereby allowing the carbon dioxide they contain to be recycled back into the Earth's air. But because it is smaller, Mars has lost too much of its internal heat to drive plate tectonics. As a result, once atmospheric carbon dioxide is fixed into carbonate minerals, it stays there.

Thus, over hundreds of millions of years, the thick carbon dioxide blanket that warmed Mars in its youthful years gradually thinned out, causing the global temperature to drop precipitously. As the atmospheric carbon dioxide thinned out through carbonate fixation, temperatures dropped to the point where the soil became an effective sorbent for the gas, sponging it right out of the air. The colder it got, the stronger the soil sorbent became, resulting in a runaway ice-box process that froze the whole planet solid.

But in the three billion years since that disaster occurred, the Sun has grown in power by some 30 percent, so that now it is believed that a further temperature increase of about ten centigrade could trigger the reverse process. That is, if we today can somehow artificially induce a certain amount of positive global warming, the increased temperature itself would cause some of the carbon dioxide currently sorbed in the soil to outgas, which would thicken the atmosphere and add to the greenhouse effect.

Central to understanding how powerfully this could drive warming is the concept of positive feedback, a phenomenon that occurs when the output of a system enhances what is input to the system. For a

Mars greenhouse effect, we find positive feedback in the relationship between atmospheric pressure—its thickness—and atmospheric temperature. Heating Mars will release carbon dioxide from the polar caps and from Martian regolith. The liberated carbon dioxide thickens the atmosphere and boosts its ability to trap heat. Trapping heat increases the surface temperature and, therefore, the amount of carbon dioxide that can be liberated from the ice caps and Martian regolith. And that is the key to terraforming Mars—the warmer it gets, the thicker the atmosphere becomes; and the thicker the atmosphere becomes, the warmer it gets.

To understand how this works, look at Figure 11.1, which shows the dynamics of the Martian regolith/atmosphere carbon dioxide greenhouse system. The curve marked with the little squares shows the average global temperature as a function of the carbon dioxide atmosphere's pressure. Here we see the predicted results of the greenhouse effect. The thicker the atmosphere, the warmer the planet gets. The line with the diamonds shows the soil vapor pressure as a function of the global temperature. The warmer the planet gets, the more carbon dioxide vaporizes from the poles and outgases from the regolith.

Note the two points, A and B, where the curves cross. Each is an equilibrium point where Mars's mean atmospheric pressure and average temperature (given in degrees Kelvin; to translate into Centigrade, just subtract 273. 273 K = 0° C) given by these two curves are mutually consistent. The arrows show the direction the system will move if it is displaced from equilibrium. A is a stable equilibrium, while B is unstable. This can be seen by examining the dynamics of the system wherever the two curves do not coincide. Whenever the temperature curve lies above the vapor pressure curve, the system will move to the right, or toward increased temperature and pressure; this would represent a runaway greenhouse effect. Whenever the temperature curve lies below the pressure curve, the system will move to the left, or toward decreased temperatures and pressure; this would represent a

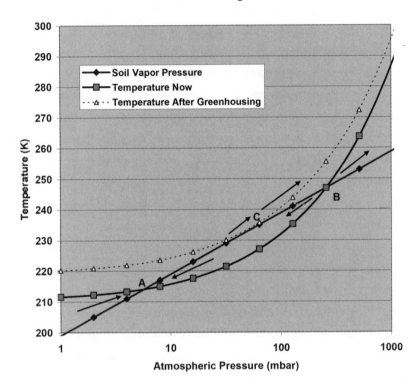

FIGURE 11.1. *To terraform Mars, we need to lift the temperature curve above the soil vapor pressure line. Once points A and B meet at C, the system will have no stable equilibrium and a runaway greenhouse effect will result.*

runaway "icebox effect." Mars today is at point A, with 6 millibars of pressure and an average global temperature of about 215 K.

Now consider what would happen if settlers artificially increased the average temperature of Mars by 8 degrees Kelvin. The results of such a change are shown by the dashed curve marked with little triangles. As the temperature is increased, the solid temperature curve would move upward to where the dashed curve is marked, causing points A and B to move toward each other until they meet at point C. It will be observed that the average global temperature at C is 230 K, which is 15 degrees warmer than the 215 K temperature we started with at A, so the effect of the original artificial 8-degree temperature rise input to the system has been substantially amplified by positive feedback. But more important, the new temperature curve is above the pressure curve everywhere, so point C is an unstable equilibrium. Once this is reached, the result would be a runaway greenhouse effect that would cause the entire available reservoir of polar and regolith carbon dioxide to evaporate and outgas, driving temperatures and pressures upward and outward along the dashed curve. As soon as the pressure has moved out past the current unstable equilibrium location, the roughly 200 mbar point B, Mars will be in a runaway greenhouse condition even without artificial heating, so even if we stop the heating activity later, the atmosphere will remain in place.

Between 6 mb of carbon dioxide in the atmosphere, close to 100 mb in the poles, and perhaps about 400 mb in the regolith, there is probably enough available carbon dioxide to create an atmosphere with around half the surface pressure of Earth.[1] Looking at the data on the graph, we can see that under such conditions the average global temperatures might rise to about 275 K—in other words, slightly above the freezing point of water, with tropical equatorial and summertime temperate zone climates being considerably warmer.

That's good enough to create a living planet.

How Fast Would the Atmosphere Come Out of the Regolith?

But how fast could this be done? The carbon dioxide obtained from the polar cap will come off quickly, but forcing out adsorbed carbon dioxide from regolith at significant depth might take some time. For terraforming to be of practical interest (particularly to investors seeking a high rate of appreciation on their real-estate claims). the rate at which all this occurs is important. After all, if it takes ten million years for a substantial amount of gas to come out of the regolith, the fact that it comes out eventually would be rather academic.

The rate that gas comes out of the regolith will be in direct proportion to the rate at which a temperature increase that we create on the surface of Mars can penetrate the ground. The thermal conductivity of Martian regolith is probably close to that of dry soil on Earth, with a little bit of ice mixed in. The rate at which heat will spread through such a medium is governed by the process of thermal conduction, whose equations predict that the time a temperature rise needs to travel a given distance through a medium is proportional to the square of the distance. Based on terrestrial analogs, this is likely to be about 16 square meters per year. Martian regolith, which has an average density of about 2.5 metric tons per cubic meter, includes a lot of clay-like minerals, and it is reasonable to assume that it is saturated with about 5 percent carbon dioxide down to considerable depth.[2] If this is true, we would have to force out carbon dioxide held in regolith—outgas it—down to a depth of 100 meters to produce a 500 millibar (0.5 bar, or half Earth sea-level) pressure on Mars. So, let's say we induced a sustained artificial temperature rise at the surface of 10 Kelvin, good enough to outgas a significant fraction of what is in the regolith. This temperature rise would then travel down into the ground. The rate at which this would occur is shown in Figure 11.2.

You can see that while it takes a very long time to reach significant depths, modest depths can be reached rather quickly. So, while it might take two hundred years to penetrate a hundred meters to get about 300 millibars out of the regolith, the 100 millibars can be gotten out in just a few decades.

Once significant regions of Mars rise above the freezing point of water on at least a seasonal basis, the large amounts of water contained in glaciers or frozen into the regolith as permafrost would begin to melt, and eventually flow out into the dry riverbeds of Mars. There

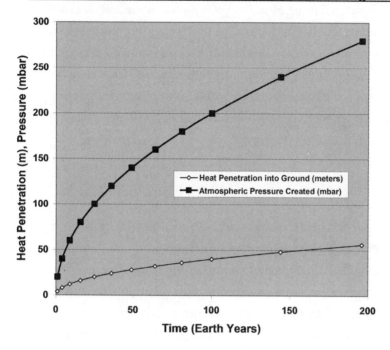

Rate of Outgassing of Atmosphere from the Martian Regolith

FIGURE 11.2. *After the surface warms, it will take time for the heat to soak into the ground. In a hundred years, the warming will reach a depth of forty meters, releasing two hundred millibars of carbon dioxide into the atmosphere.*

will also be rain and snow, which will spread water widely and destroy much of the perchlorates that contaminate Mars's soil, releasing some oxygen in the process. Water vapor is also a very effective greenhouse gas, and since the vapor pressure of water on Mars would rise enormously under such circumstances, the reappearance of liquid water on the Martian surface would add to the avalanche of self-accelerating effects all contributing toward the rapid warming of Mars. The seasonal availability of liquid water will also allow us to spread bacteria and green plants. Natural bacteria exist that produce methane and ammonia, and artificial species can probably be engineered that will be able to produce even more effective warming gases. These will augment the greenhouse effect and protect the planet against solar ultraviolet radiation. Simultaneously, photosynthetic microorganisms and plants will begin the process of oxygenating the atmosphere.

So, in short, science says that if they can start the process by artificially raising their planet's temperature by 10° C or so, Martian settlers will be able to make their world come alive. Once a strong initial warming shove is given, climate change will self-accelerate. How might that shove be delivered?

Producing Halocarbons on Mars

Ideas for initiating global warming on Mars break down into two general categories: violent and nonviolent.

Violent methods include bombarding Mars's polar cap with energetic objects, such as hydrogen bombs or impacting asteroids, with the object of vaporing the cap's large carbon dioxide reservoir. This would cause the planet's surface temperature to rise nearly instantly, possibly enough for the desired runaway greenhouse effect to take off.

There are major problems with these ideas. Earth's current arsenal of hydrogen bombs contains insufficient energy to vaporize the Martian south pole, and, in any case, the nations that possess such weapons have other priorities for their use. To make this concept work, the Martians would need to manufacture millions of more multi-megaton H-bombs. Such an arms program would cost an immense amount of money, and doubtless raise serious concerns among Martian city-states and between Mars and the home planet. Furthermore, once they were dropped on the pole, the radioactive fallout would spread planetwide, mixing with Martian dust and water to create a health hazard that would last for centuries.

Asteroids made of frozen volatiles (a.k.a. "iceteroids") can be found in the Kuiper Belt and Oort Cloud. Because they are situated far from the Sun, they circle it very slowly, so that tiny changes in their velocity can effect great changes in their orbital parameters, shifting them into highly elliptical orbits that visit the inner solar system. This happens naturally from time to time when a passing star moves through our Oort Cloud, destabilizing thousands of objects into cometary orbits. Depending on how far out they are, these can then take centuries to millennia to reach the inner solar system where they can display themselves as magnificent comets to amuse the public or impact planets to cause mass extinctions.

Using the volatile content of such objects as propellant for use in nuclear thermal rockets,[3] astronautical engineers working in the Kuiper Belt could destabilize the orbit of such iceteroids and send them into the inner solar system as very high energy projectiles. For example, if an iceteroid with a mass of 10 billion tons were orbiting the Sun in the inner Kuiper Belt at 40 AU, it would only take a ΔV of 200 m/s to destabilize it into an orbit that would intersect that of Neptune, which could provide a gravity assist that would send it on to Mars. Three 5,000 MW nuclear reactors powered by either fission or fusion could generate the required impulse with about ten years of thrusting. About 5 percent of the iceteroid's volatile content would be used as propellant, leaving the remaining 9.5 billion tons to slam into Mars at a velocity on the order of 40 km/s about a century later. The energy released on impact would be about 8 trillion GJ, roughly equal to 2 million one-megaton H-bombs. That would be enough to vaporize the south pole, and provided that the nuclear reactors were removed prior to impact, there would be no radioactive fallout at all.

The problem is that such an incoming object would create serious range safety concerns for the Martians, as, if the aim was off by just a little bit, it could devastate the planet. It might well devastate the planet even if the aim was accurate. So, I doubt very much that the Martians would go for it.

This leaves nonviolent methods. These include using orbital mirrors to vaporize the polar cap or setting up factories to produce artificial greenhouse gases.

As I discuss in my book *The Case for Mars*,[4] building an orbital mirror composed of solar sail material capable of vaporizing the pole would be possible in principle. Based upon the total amount of solar energy required to raise the temperature of a given area a certain number of degrees above the polar value of 150° Kelvin, it turns out that a space-based mirror with a radius of 125 kilometers could reflect enough sunlight to raise the entire area south of 70° south latitude by 5° Kelvin—more than enough to vaporize the cap. If made of solar sail–type aluminized mylar material with a density of 4 tons per square kilometer (about 4 microns thick), such a sail would have a mass of 200,000 tons. Many ships of this size are currently sailing the Earth's oceans. Thus, while this is too large to consider launching from Earth, if space-based manufacturing techniques are available, its construction in space out of asteroidal or Martian moon material is a serious option. The total amount of energy required to process the materials for such a reflector would be about 120 MWe-years, which could be readily provided by a set of 5 MWe nuclear reactors such as might be used in piloted nuclear electric propulsion (NEP) spacecraft. Interestingly, if stationed near Mars, such a device would not have to orbit the planet. Rather, solar light pressure could be made to balance the planet's gravity, allowing the mirror to hover as a "statite"[5] with its power output trained constantly at the polar region. For the sail density assumed, the required operating altitude would be 214,000 kilometers.

The problem with this concept, however, is that it would only warm the polar region. If the amount of carbon dioxide in the cap were insufficient to set off the runaway greenhouse effect, it would fail to get the job done.

That leaves the production of artificial greenhouse gases as the best option for warming Mars. Martian settlers could set up factories to produce halocarbons, which are the strongest greenhouse gases known to man. The best known such gases are the chlorofluorocarbons, or "CFCs." Commonly used as refrigerants in the past, these were banned by treaty in the 1990s. This was partly because of their strong contribution to the greenhouse effect but primarily because they were blamed for the destruction of the ozone layer. However, while equally powerful as greenhousing agents, halocarbons lacking in chlorine (i.e., fluorocarbons) do not destroy ozone. The easiest such gas to make is perfluoromethane, CF_4, which also has the desirable feature of being very long-lived (stable for more than ten thousand years) because it is immune to destruction by ultraviolet light. The greenhousing effect of using CF_4 can be improved by throwing in smaller amounts of other fluorocarbon gases, such as C_2F_6 and C_3F_8, which serve to block up the gaps in the infrared spectrum that an atmospheric blanket of CF_4 and carbon dioxide alone might leave open. In Table 11.1, you can see the amount of such a fluorocarbon gas cocktail needed in Mars's atmosphere to create a given temperature rise, and the power that we would need to generate on the Martian surface to produce the required fluorocarbons over a period of twenty years. If the gases have an atmospheric lifetime of a hundred years, then approximately one-fifth the power levels shown in the table will be needed to maintain the CFC concentration after it has been built up. As you can see, the Martians are going to need to mount a substantial industrial effort to pull this off—something like two to four gigawatts (a gigawatt, GW, is one thousand megawatts, roughly the residential electric power used by a modern American city with a population of one million people) if they wish to build up such a gas blanket in a timely way. Still, given the value the terraforming project will offer to Martian settlers, it's an amount of power that could become available once the planet's population reaches ten million or so.

TABLE 11.1: Greenhousing Mars with CFCs

Induced Heating	CFC Pressure	CFC Production	Power Required
(degrees K)	(micro-bar)	(tons/hour)	(GW)
5	0.012	260	1.31
10	0.04	880	4.49
20	0.11	24,10	12.07
30	0.22	4,830	24.15
40	0.80	17,570	87.85

Oxygenating the Planet

As the planet warms, its hydrosphere will be activated. Water will melt out from the ice and permafrost, flow into the streams, rivers, and lakes, evaporate, and come down everywhere as rain and snow. The more rapidly water gets into circulation, the faster denitrifying bacteria will break down nitrate beds. This will increase the atmospheric nitrogen supply and accelerate the spread of oxygen-producing plants. . . . Activating the hydrosphere will also serve to destroy the oxidizing chemicals in the Martian regolith, thereby releasing some additional oxygen into the atmosphere in the process.

Thickening Mars air with carbon dioxide will greatly benefit human settlers because once atmospheric pressure rises above 200 mb it will no longer be necessary to wear spacesuits. Instead, those venturing outside will only need to wear warm clothing and oxygen masks, much as high altitude bomber pilots did in World War II and Mt. Everest climbers do today. Furthermore, it will become readily feasible to create city habitation domes of any size, since there will no longer be a need to have a pressure difference between the inside and the outside. The amount of available farmland will also expand enormously, since for some crops greenhouses will no longer be necessary.

But releasing enough oxygen into the air to make it breathable for humans is going to be a challenge. Bacteria and primitive plants can survive in an atmosphere without oxygen, but advanced plants require at least 1 mbar and humans need 120 mbar. While Mars does have super-oxides in its regolith that can be heated to release oxygen (and possibly nitrates that can release nitrogen), breaking down the dominant iron oxide (or mixed iron, aluminum, silicon, magnesium, etc. metal oxide) compounds to produce a substantial oxygen atmosphere would require enormous amounts of energy, about 2 million GW-years for every millibar produced—and that's just too expensive to be practical. Currently the entire human race uses about 20,000 GW of power. So, even if a power plant equal to that driving all current civilization were available, producing oxygen from regolith using industrial means would take still one hundred years per millibar.

Similar amounts of energy are required for plants to release oxygen from carbon dioxide. Plants, however, offer the advantage that once established they can propagate themselves, and draw their power from the 80 million GW of solar energy falling naturally on Mars.

The production of an oxygen atmosphere on Mars will thus break down into two phases. In the first phase, pioneering cyanobacteria and primitive plants will be employed to produce sufficient oxygen (about 1 millibar) to allow advanced plants to propagate across Mars. Once an initial supply of oxygen is available, and with a temperate climate, a thickened carbon dioxide atmosphere to supply pressure and greatly reduce the space radiation dose, and a good deal of water in circulation, plants that have been genetically engineered to tolerate Martian regoliths and to perform photosynthesis at high efficiency will be released together with their bacterial symbiotes. Assuming that global coverage could be achieved in a few decades and that such plants could be engineered to be 1 percent efficient (rather high, but not unheard of among terrestrial plants) then they would represent an equivalent oxygen-producing power source of about 800,000 GW. Using such biological systems, the required 120 millibars of oxygen needed to support humans and other advanced animals in the open could be produced in about three hundred years.

That's still too slow for most people's taste. So, Martians will be driven to engineer much more powerful artificial energy sources or still more efficient plants, or perhaps truly artificial self-replicating photosynthetic machines. Such technology could accelerate this schedule radically.

It may be noted that thermonuclear fusion power on the terawatt scale required for the acceleration of terraforming also represents the key technology for enabling piloted interstellar flight. If terraforming Mars were to produce such a spin-off, then the ultimate result of the project will be to confer upon humanity not only one new world for habitation, but myriads.

Thus, Mars could be transformed from its current dry and frozen state into a warm and wet planet capable of supporting life.

While humans might not be able to breathe the outside air straightaway, simple hardy plants could, and in fact thrive in the carbon dioxide–rich outside environment to spread rapidly across the planet's surface. Over centuries, these plants will introduce oxygen into Mars's atmosphere in increasingly breathable quantities, opening the surface to advanced plants and growing numbers of animal types. As this occurs, the carbon dioxide content of the atmosphere will be reduced, which would cause the planet to cool unless greenhouse gases are introduced capable of blocking off those sections of the infrared spectrum previously protected by carbon dioxide. But the Martians should certainly be smart enough to deal with a matter like that.

It might take a while, but eventually the day will come when the domed tents will no longer be necessary, and the descendants of the first settlers will be able to throw away their oxygen masks to inhale the glorious scent of the towering evergreen forests of Mars.

It's a magnificent vision.

But how will it be paid for?

FIGURE 11.3.
Eventually the day will come when city domes will no longer be necessary. (Credit: Hewlett Packard Mars Home Planet)

Terraforming for Profit

Martians will ardently support the terraforming project because it stands to make their land much more valuable. Martian land is nearly worthless today, but priced at $1,000/acre, the surface of the Red Planet would be worth about $40 trillion. The fact that most of the effects of the terraforming effort won't occur for a century or more after it is begun won't be a showstopper. Once everyone knows that extraordinary physical improvements are on the way, market values for real estate in general will rapidly rise, while the price of selected properties will soar.

As an important example of the above, consider the potential sales value of future beachfront property. On Earth, properties that front bodies of water sell for a high premium, and the same will likely be true on Mars once the terraforming program brings back into being the planet's many ancient ponds, lakes, rivers, seas, and oceans. Good opportunities to make super profits might be found in the field of hydroelectric power. On Earth, much of the expense involved in building a large hydroelectric dam is a consequence of the huge effort required to take a flowing river and diverting it away from the dam site so that the structure can be built, after which the river needs to be diverted back. On Mars, however, there won't be that problem, as the rivers that will drive its hydroelectric installations do not yet exist. Thus, it could be possible to build hydroelectric dams on the Red Planet at much lower cost than was ever possible on Earth. Furthermore, given the extreme variations in topography that Mars affords, with mountains twenty-seven kilometers high and canyons five kilometers deep, it's clear that the hydroelectric potential of the Red Planet is immense. The future revenues of Mars's hydroelectric industry could well be astronomical.

It's true that on Earth we know precisely where the shore of a lake or an ocean is, whereas we don't know exactly how high sea level will rise on Mars, so a property that might be a prime future beachfront value with equal likelihood could end up far from shore, or worse yet, underwater. This will make such investments a gamble since *any* property on the slope of a basin or valley could readily be marketed as future beachfront. Similar scams, backed by suitably compensated expert opinions, could be run by those seeking investments in hydroelectric projects. As occurred in colonial and frontier America, the best promoted of such ventures will likely sell billions of dollars' worth of shares to Earthside investors. Some of these will no doubt make terrific profits themselves unloading the paper on others, and those others on still others, for at least a century, as the terraforming program advances and the prospect of producing some hydroelectric power becomes ever more tangible, at least in theory. The central concept to keep in mind is this: *Property titles are not for using; they're for buying and selling.* It worked for the crypto market. It could work for Martian land speculation just as well.

I've been kidding around a bit here, but there is a basic point to be made. As occurred in conjunction with the building of the railroads to the American West, the terraforming of Mars will radically affect real

FIGURE 11.4.
The terraforming program will transform barren desert into prospective high-value tropical beachfront property. (Credit: Michael Carroll)

estate values. This will put into play a great deal of speculative money. There will always be people willing to gamble for the chance to make a big score. As with all the grand projects of the past, both the noble and base sides of human nature will play their parts. Fortunes will be made, and fortunes will be lost. But the job will get done.

Terraforming for God

> *Witness this new-made World, another Heav'n*
> *From Heaven Gate not farr, founded in view*
> *On the clear Hyaline, the Glassie Sea;*
> *Of amplitude almost immense, with Starr's*
> *Numerous, and every Starr perhaps a World*
> *Of destined habitation.*
>
> —JOHN MILTON, *Paradise Lost*

The practical value of terraforming their planet will be clear to all Martians. It will make their land vastly more valuable and their businesses far more profitable. Some cities might seek to benefit from the project without doing their share of the work. But if the planet is to be terraformed, most of the leading city-states will need to play their part. As diverse as they may otherwise be, this common goal will tend to bring Martians together.

Terraforming will unite Mars, and not just as a matter of common endeavor for material gain. Terraforming will unite Mars in spirit.

Over the years I have engaged in several debates with people who claim that it would be "unethical" for humans to terraform Mars because Mars supposedly has a "right" to be as it is, without humans imposing their will upon it. According to such people, the terraforming of Mars would just be another version of imperialism, and hence evil.

I believe that such views are nonsensical. The imposition of the will of advanced Western societies on indigenous peoples, such as native Americans, for example, was evil because the natives were people. They were not rocks or microbes, and to compare such entities to American Indians is to greatly degrade the latter. People have rights. Rocks and microbes do not. To claim that humans do not have the right to alter Mars because Mars has the right to be unaltered is as nutty as claiming that Michelangelo was committing criminal mutilation of marble blocks by chiseling them into statues. Those advancing such arguments are not really proposing to defend the rights of rocks. They are proposing to destroy the rights of humans.

Nevertheless, such deeply antihuman views have become fashionable among many of those who consider themselves the enlightened. "The earth has cancer, and the cancer is man," proclaimed the elite Club of Rome in one its manifestos. This is a very dangerous point of view. One does not give freedom to a cancer.

The Martians will have no patience for such nonsense. This will be so for two reasons. In the first place, the governing ethos of any successful Martian city-state must and will be one of cheerful Promethean pragmatism. But more than that, they will expose antihumanism's total absurdity in the grand light of their noble project to bring a dead planet to life.

The discussion of humanity's potential to terraform Mars brings us to a fundamental question. Are we first-class citizens of the cosmos, or are we beings of lesser order? Kepler proved that the laws of the heavens were understandable by the human mind. The first astronauts to reach Mars will prove that the worlds of the heavens are accessible to human life. But if we can terraform Mars, it will show that the worlds of the heavens themselves are subject to the human intelligent will.

By continuing the work of creation, we can make Mars a second home for life, all life; not only humans, nor even just for "the fish of the sea, . . . the fowl of the air, and every living thing that moveth upon the Earth," but for a plenitude of species yet unborn. New worlds invite new forms, and in the novel habitats that a terraformed Mars would provide, life brought from Earth could go forth and multiply into realms of diversity yet unknown.

This is the wondrous heritage that the Martians will leave for all humanity for all time—not only a new world for life and civilization, but an example of what men and women of intelligence, daring, and vision can accomplish when acting upon their highest ideals. Gods we'll never be. But the humanity that terraforms Mars will have shown that humans are more than just animals, that we are in fact creatures who carry a unique spark that is worthy of respect. No one will be able to look at the new Mars without feeling prouder to be human. No one will be able to hear its story without being inspired to rise to the tasks that will lie ahead among the stars.

12

THE CAUSE OF
ALL NATIONS

To the frontier the American intellect owes its striking characteristics. That coarseness and strength combined with acuteness and inquisitiveness; that practical, inventive turn of mind, quick to find expedients; that masterful grasp of material things, lacking in the artistic but powerful to effect great ends; that restless, nervous energy; that dominant individualism, working for good and for evil, and withal that buoyancy and exuberance which comes with freedom—these are traits of the frontier, or traits called out elsewhere because of the existence of the frontier.

—FREDERICK JACKSON TURNER,
"The Significance of the Frontier in American History," 1893[1]

We shall never surrender, and even if, which I do not for a moment believe, this Island or a large part of it were subjugated and starving, then our Empire beyond the seas, armed and guarded by the British Fleet, would carry on the struggle, until, in God's good time, the New World, with all its power and might, steps forth to the rescue and the liberation of the old.

—WINSTON CHURCHILL,
"We shall fight on the beaches" speech, June 4, 1940[2]

AMERICA SAVED CIVILIZATION.

In 1492 the embryonic version of the entity we now call Western civilization was a collection of small countries clinging on to the Atlantic-Mediterranean fringe of the Eurasian continent calling itself

"Christendom." Besieged by an expansionist Islamic superpower on its border, and dwarfed in size, population, and wealth by much vaster empires farther east, Christendom nevertheless held unique potential. While it may have been materially poor, Christendom was intellectually rich. From its combination of Hellenistic and Judeo-Christian heritages it had derived the idea of the human mind as the image of God,[3] possessing an intrinsic ability to distinguish right from wrong, justice from injustice, and truth from untruth. From this idea came both the Christian notion of the *conscience*, and ultimately the related concept of *science*. That's right. Science is a Western *religious* concept. It is founded on the *faith* that the universe needs to make sense to the human mind. As the Renaissance scientist Johannes Kepler, the discoverer of the laws of planetary motion, from which Newton's laws and ultimately all classical physics were subsequently derived, put it: "Geometry is one and eternal, a reflection out of the mind of God. That mankind shares in it is one reason to call man the image of God."

As the image, or children, of God, humans have intrinsic worth, and should be freed from the slavery of muscle toil. God wants us to be free to continue the work of creation. So, those who help set us free through invention are serving the divine cause.

The fact that we have a conscience gives us not only the right, but the obligation, to judge the justice or injustice of statute laws using human reason. So, there must be a higher law, or *natural law*—derivable by reason—that sets the standard of justice against which the laws of governments can be judged. If God wants us to be free, then natural law must dictate that humans have *natural rights*.

These ideas were all implicit in Christendom. But before some of its descendants could sit down and spell out their meaning with "We hold these truths to be self-evident . . ." they needed a place to do it.

America gave them that place—a place in which there were no established ruling institutions, an improvisational theater big enough to welcome all comers with no parts assigned. On such a stage, the players are not limited to the conventional role of actors—they become playwrights and directors as well. The unleashing of creative talent that such a novel situation allows is not only a great deal of fun for those lucky enough to be involved, but it also changes the opinion of the spectators as to the capabilities of actors in general. People who had no role in the old society could define their role in the new. People who did not "fit in" in the Old World could discover and demonstrate that, far from being worthless, they were invaluable in the New, whether they journeyed there or not.

The New World offered by America destroyed the basis of aristocracy and created the basis of democracy. It allowed the development of diversity by allowing escape from those institutions that imposed uniformity. It destroyed a closed intellectual world by importing unsanctioned data and experience. It allowed progress by escaping the hold of those institutions whose continued rule required continued stagnation, and it drove progress by defining a situation in which innovation to maximize the capabilities of the limited population available was desperately needed. It raised the dignity of workers by raising the price of labor and by demonstrating for all to see that human beings can be the creators of their world. In America, from colonial times through the nineteenth century when cities were rapidly being built, people understood that America was not something one simply lived in—it was a place one helped build. People were not simply inhabitants of their world. They were makers of it.

The interaction with American inventive culture drove technological progress in Europe, inspiring emulation from gadgeteers there and setting off a transatlantic race for patents. It also set a new standard for government—democracy and human rights. And when the forces of reaction threatened to throw Europe back into the Dark Ages, the New World did indeed, as Churchill prayed in the desperate days of June 1940, come to the rescue and liberation of the Old.[4]

Nineteenth-century European freedom fighters understood that America was their main hope. That is why when the struggle between liberty and slavery as the basis for the organization of society was put to the test within America during our Civil War they rallied in support of the Union as "The Cause of All Nations."[5]

One ardent supporter of the Union cause was the French sculptor Frederick Bartholdi. It was during the war that Bartholdi conceived of his most famous project, the Statue of Liberty, which was eventually completed and erected in New York Harbor in 1886. Most Americans today view that statue through the lens of Emma Lazarus's poem, which poses Liberty as welcoming immigrants to America with the words "bring me your huddled masses yearning to breathe free." Fair enough. But in Bartholdi's mind, Liberty had a much more aggressive message. He entitled his work "Liberty Shining Her Light to the World." Liberty is not facing east just to welcome immigrants coming into New York Harbor. She faces east to send a message all the way to Europe: Arise and be free!

But Liberty was not the only statue that Bartholdi created to celebrate America. He also made one of Christopher Columbus. That one, made of silver, was displayed at the 1892 Chicago Exhibition on the four hundredth anniversary of the explorer's historic voyage. Not meant for permanent exhibition, that statue was later melted down to recover its very expensive contents. But a copy of it, in bronze, was put on display with great ceremony in Providence, Rhode Island, in 1893. There it stood for 127 years, until, under pressure from demonstrators claiming to be more "woke" than prior generations, the Providence City Council agreed in 2020 to take it down.[6]

In canceling Columbus, countercultural radicals are attempting to revise history to say that America was born under the mark of Cain. Furthermore, since the entire American enterprise was criminal, so are the virtues that enabled its creation.

Bartholdi's statue of Columbus captures him as the epitome of the spirit of exploration—the spirit of the frontier, boldly venturing to challenge the unknown, human reason unfazed by the terrors of nature—immanent in America from the moment of its birth. Furthermore, he is holding a globe, showing that his courage is based on his faith in science. Like Odysseus, the hero of the founding epic of Western civilization, he uses the gifts of Athena, the goddess of practical wisdom, to take on Poseidon, the god of the sea.[7] It's a very powerful statement.

That is why it was removed.

Which raises the question of the future fate of his other statue.

FIGURE 12.1. *Two statues by Frederick Bartholdi. Left: Liberty Shining Her Light to the World (1886). Right: Christopher Columbus (1892). Columbus was removed in 2020. Liberty still stands, for now. (Credit: Wikipedia Commons)*

The term "decadent" is sometimes understood to mean a pornographic hedonistic lifestyle devoted to the pursuit of pleasure. That certainly can be part of it, as exemplified in late Rome. But more broadly, a society can be said to become decadent when it abandons the principles that gave birth to it and allowed it to grow and prosper. The Roman aristocracy became decadent when its ideal life was no longer one that achieved fame for great public service, but one that could enjoy the most fabulous parties and orgies, or alternatively, steal the most wealth for oneself.[8] Rome's commoners became decadent when they lost their farms and transitioned from being citizen defenders of the republic to an urban mob willing to sell its votes for bread and circuses. Without public-spirited leaders or citizens, the Roman republic fell into empire, and then decaying further, collapsed altogether.

As it abandons the frontier spirit, America also threatens to become decadent. Currently we see around us an ever more apparent loss of vigor of American society: increasing fixity of the power structure and bureaucratization of all levels of society; impotence of political institutions to carry off great projects; the proliferation of regulations affecting all aspects of public, private, and commercial life; the spread of irrationalism; the banalization of popular culture; the loss of willingness by individuals to take risks, to fend for themselves or think for themselves; economic stagnation and decline; the deceleration of the rate of technological innovation. Everywhere you look, the writing is on the wall.

Without a frontier from which to breathe life, the spirit that gave rise to the progressive humanistic culture that America has offered to the world for the past several centuries is fading. In rejecting the values that allowed for America's creation, we are becoming less than the people we used to be.

If we abandon the spirit of exploration, how long will we sustain that necessary for liberty?

The issue is not just one of national loss—human progress needs a vanguard, and no replacement is in sight.

The creation of a new frontier thus presents itself as America's and humanity's greatest social need. Nothing is more important: Apply what palliatives you will, without new frontier societies to stimulate it, the entire global civilization based upon Western values of humanism, reason, science, conscience, self-reliance, individual liberty, and progress could ultimately die.

Humanity Needs New Nations

"Everything has tended to regenerate them; new laws, a new mode of living, a new social system; here they are become men."

—JEAN DE CREVECOEUR,
Letters from an American Farmer, 1782

Consider the probable fate of humanity in the twenty-first century under two conditions: with Martian frontier city-states and without them.

In the twenty-first century, without a Martian frontier, there is no question that human cultural diversity will decline severely. Already, in the early twenty-first century, advanced communication and transportation technologies have eroded the healthy diversity of human cultures on Earth. As technology allows us to come closer together, so we come to be more alike. Finding a McDonald's in Beijing, country and western music in Tokyo, or a Nikola Jokic T-shirt on the back of an Amazon native is no longer a great surprise.

Bringing together diverse cultures can be healthy, as it sometimes results in fusions that produce flowerings in the arts. It can also result in very unpleasant increases in ethnic tensions. But however the energy released in the cultural merger is expended in the short term, the important thing in the long term is that it is expended. An analogy to cultural homogenization is that of connecting a wire between the terminals of a battery. A lot of heat can be generated for a while, but when all the potentials have been leveled, a condition of maximum entropy is reached, and the battery is dead. Consider what happened after the diverse vibrant cultures of the ancient Mediterranean world were merged into the Roman Empire. The golden age produced by unification is frequently followed by stagnation and decline.

The tendency toward cultural homogenization on Earth can only accelerate in the twenty-first century. Furthermore, because of rapid communication and transportation technologies "shorting out" terrestrial intercultural potential barriers, it will become increasingly impossible to obtain the degree of separation required to develop new and different cultures on Earth. If the Martian frontier is opened, however, this same process of technological advance will also enable us to establish many new, distinct, and dynamic branches of human culture on Mars and eventually more on worlds beyond. The precious diversity of humanity can thus be preserved on a broader field, but only on a broader field. One world will be just too small a domain to allow the preservation and continued generation of the diversity needed not just to keep life interesting, but to assure the survival of the human race.

Without the opening of a new frontier on Mars, continued Western civilization faces the risk of *technological* stagnation. To some this may appear to be an outrageous statement, as the present age is frequently cited as one of technological wonders. In fact, however, the rate of progress within our society has been decreasing and at an alarming rate. To see this, it is only necessary to step back and compare the changes that have occurred in the past forty years with those that occurred in the preceding forty years and the forty years before that.

Between 1903 and 1943, the world was revolutionized: Cities were electrified; telephones and broadcast radio became common; talking motion pictures appeared; automobiles became practical and commonplace; and aviation progressed from the Wright Flyer to the DC-3 and the Spitfire. Between 1943 and 1983, the world changed again, with the introduction of communication satellites and interplanetary spacecraft, computers, television, antibiotics, nuclear power, Atlas, Titan, and Saturn rockets, Boeing 737s and SR-71s. Compared to these changes, the technological innovations from 1983 to the present seem insignificant. Immense changes should have occurred during this period but did not. Had we been following the previous eighty years' technological trajectory, we today would have flying cars, maglev (magnetic levitation) trains, jetpacks, breeder reactors, fusion reactors, hypersonic intercontinental travel, reliable and inexpensive transportation to Earth orbit, undersea cities, true robots, open-sea mariculture, and human settlements on the Moon and Mars. Instead, today we see important technological developments, such as nuclear power and biotechnology, being blocked or enmeshed in political controversy. We are slowing down.

Now, consider a nascent Martian civilization: Its future will depend critically upon the progress of science and technology. Just as the inventions produced by the "Yankee ingenuity" of frontier America were a powerful driving force on worldwide human progress in the nineteenth century, so the "Martian ingenuity" born in a culture that puts the utmost premium on intelligence, practical education, and the determination required to make real contributions will make much more than its fair share of the scientific and technological breakthroughs that will dramatically advance the human condition in the twenty-first.

A prime example of the Martian frontier driving new technology will undoubtedly be found in the arena of energy production. As on Earth, an ample supply of energy will be crucial to the success of Mars settlements. The Red Planet does have one major energy resource that we currently know about: deuterium, which can be used as the fuel in nearly waste-free thermonuclear fusion reactors. Earth has large amounts of deuterium, too, but with all the existing investments in other, more polluting forms of energy production, the research that would make possible practical fusion power reactors has been allowed to stagnate. The Martian colonists are certain to be much more determined to get fusion online, and in doing so will massively benefit the mother planet as well.

The parallel between the Martian frontier and that of nineteenth-century America as technology drivers is, if anything, vastly understated. America drove technological progress in the last century because its western frontier created a perpetual labor shortage back East, thus forcing the development of labor-saving machinery and providing a strong incentive for improvement of public education so that the skills of the limited labor force available could be maximized. This condition no longer holds true in America. In fact, far from prizing each additional citizen, anti-immigrant attitudes are on the rise, and a vast "service sector" of bureaucrats and menials is being created to absorb the energies of those parts of the population whose participation in the productive parts of the economy is no longer needed. Thus, in the early twenty-first century, each additional citizen is and will be regarded as a burden.

On twenty-first-century Mars, on the other hand, conditions of labor shortage will apply with a vengeance. Indeed, it can be safely said that no commodity on Mars will be more precious, more highly valued, and more dearly paid for than human labor time. Workers on Mars will be paid more and treated better than their counterparts on Earth. Just as the example of nineteenth-century America changed the way the common man was regarded and treated in Europe, so the impact of progressive Martian social conditions may be felt on Earth as well as on Mars. A new standard may be set for a higher form of humanist civilization on Mars, and, viewing it from afar, the citizens of Earth will rightly demand nothing less for themselves.

The frontier drove the development of democracy in America by creating a self-reliant population that insisted on the right to self-government. It is doubtful that democracy can persist without such people. True, the trappings of democracy exist in abundance in

America today, but meaningful public participation in the process is deeply wanting. Consider that no representative of a new political party has been elected president of the United States since 1860. Likewise, neighborhood political clubs and ward structures that once allowed citizen participation in party deliberations have vanished. And with a reelection rate of 95 percent, the US Congress is hardly susceptible to the people's will. Regardless of the will of Congress, the real laws, covering ever broader areas of economic and social life, are increasingly being made by a plethora of regulatory agencies whose officials do not even pretend to have been elected by anyone.

Democracy in America and elsewhere in Western civilization needs a shot in the arm. That boost can only come from the example of a frontier people whose civilization incorporates the ethos that breathed the spirit into democracy in America in the first place. As Americans showed Europe in the last century, so in the next the Martians can show Earth the path away from oligarchy.

A House Divided?

"A house divided against itself cannot stand. I believe this government cannot endure permanently half slave and half free. I do not expect the Union to be dissolved—I do not expect the house to fall—but I do expect it will cease to be divided. It will become all one thing or all the other. Either the opponents of slavery will arrest the further spread of it and place it where the public mind shall rest in the belief that it is in the course of ultimate extinction; or its advocates will push it forward, till it shall become lawful in all the States, old as well as new—North as well as South."

—ABRAHAM LINCOLN, June 16, 1858

Mars cities will face a severe labor shortage. I believe this will drive them toward creating a culture of invention, much as happened in the early United States.

But, in fact, there were two radically different ways that youthful America attempted to deal with its labor shortage. One indeed was to use technology to multiply the productivity of labor, thereby enabling itself to pay the high wages that a tight free labor market demands. This created a vibrant society based on technological progress sustained by freedom and widespread education. By so doing, it offered opportunities that excited the hopes and dreams of the common folk of the Old World, to a degree that motivated millions of them to abandon their homes to move here at great cost and risk to have a chance to take part.

This was the solution adopted by the North.

The South, however, took a very different path. There are two ways to handle a labor shortage. You can offer high pay to attract workers. Or you can use force to compel people to work for you under whatever conditions you dictate.

That was the solution adopted by the South.

Slave labor societies do not require advanced technology. They can't invent it because they cannot tolerate educated workers, and slaves have little motivation to make their masters richer by suggesting improvements in any case. Slave societies even have great difficulty employing technology invented by others. You can't trust a slave with an expensive machine, or even a horse, because he hates you, and it's amazing how many accidents happen when property is entrusted to people who don't care for its owners. Free labor invents machines. Slave labor can't even use them.

The masters in a slave society can live a life of ease, even as the force they use to sustain their elegant lifestyles brutalizes their character. Power corrupts, and the total power over his slaves that a slave master enjoyed provided unlimited temptation to indulge in every form of cruelty and moral degradation.

Slave labor threatened the less-skilled free workers of the North with competition that would degrade them to the status of the "white trash" of the South. The gross immorality of the slave masters offended the sensibilities and consciences of the educated classes of the North, and their growing undisguised contempt outraged the dignity of the aristocracy of the South.

Many efforts to maintain unity through compromise were made, but the two ways of life were fundamentally incompatible. Eventually it came to blows.

FIGURE 12.2. *The issue of the American House Divided is settled. Lee surrenders to Grant at Appomattox, April 9, 1865. Eyewitness drawing by Alfred R. Waud. (Credit: Alfred R. Waud, Creative Commons)*

While reactionaries have continued to wage a rearguard battle down to the present day, the issue of the House Divided in America was decisively settled in a fundamental way by the surrender of the armed forces for the slavocracy at Appomattox Courthouse on April 9, 1865.

Nevertheless, the House Divided issue has continued to bedevil humanity. In her brilliant 1943 book, *The Discovery of Freedom*[9], Rose Wilder Lane (the daughter of *Little House on the Prairie* author Laura

Ingalls Wilder) quoted Lincoln's 1858 speech and pointedly used it to describe the then-ongoing fight against fascism as a global civil war, with the entire human race constituting the House Divided in question. While that war was brought to a victorious conclusion, our House here on Earth remains divided.

Could Mars turn out to be a House Divided as well?

It could, but only for a while.

If Earth remains a House Divided, it is possible that an unfree country, such as China, could establish its own Mars cities and populate them with conscripts transported to the Red Planet on an involuntary basis. But such cities could not be centers of invention. Such cities could not draw talent from around the world. Consequently, both their growth and prosperity would be limited.

There is a reason why the painting of the meeting at Appomattox shows Lee surrendering to Grant, and not the reverse. The inventive North created industries that dwarfed those of the South. It also invented military systems, such as Gatling guns, all-steel steam-powered monitor warships, reconnaissance balloons, balloon aircraft carriers, and telegraph and railroad networks far superior to any the South could muster. And it was able to draw talent from around the world: 40 percent of the Union Army were immigrants or first-generation Americans, and another 10 percent were black. In contrast, only 3 percent of the Confederacy's armed forces were immigrants, and it was able to draw no support from its own large African American population whatsoever. The result was inevitable.

The House Divided issue of Rose Wilder Lane's time was decided in much the same way. Despite the far superior professional training of the German armed forces and the fanatical courage of their Japanese militarist allies, the Axis powers were utterly crushed by the technological and industrial virtuosity of the Anglo-American alliance—with much of the talent involved coming from emigrants fleeing lands the Nazis had overrun to lend their genius to the Allied cause.[10]

Subsequently, in the Cold War, victory went to the side of the House Divided whose freedom enabled much greater technological creativity.

So, I believe, it shall be again in the current struggle pitting the West against the Russia-Iran-China Axis, and so shall it certainly be on Mars, a world where technological inventiveness will be even more supremely important than ever has been the case on Earth.

And then, should Earth ever find itself a House Divided again, there will be another New World, conceived in liberty, ready, willing, and able to come to the rescue of the Old.

The Threat to Civilization

Human civilization currently faces many serious dangers. The most immediate catastrophic threat, however, does not come from environmental degradation, resource depletion, or even asteroidal impact. It comes from *bad ideas.*

Ideas have consequences. Bad ideas can have really bad consequences.

The worst idea that has ever been is that the total amount of potential resources is fixed. It is a catastrophic idea because it sets all against all.

Currently, such limited-resource views are quite fashionable not only among futurists, but much of the body politic. But if they prevail, then human freedoms must be curtailed. Furthermore, world war and genocide would be inevitable, for if the belief persists that there is only so much to go around, then the haves and the want-to-haves are going to have to duke it out, the only question being when.

This is not an academic question. The twentieth century was one of unprecedented plenty. Yet it saw tens of millions of people slaughtered in the name of struggle for existence that was entirely fictitious. The results of similar thinking in the twenty-first could be far worse.

The logic of the limited resource concept leads down an ever more infernal path to the worst evils imaginable. Basically, it goes as follows:

Resources are limited. Therefore:

Human aspirations must be crushed.

So, some authority must be empowered to do the crushing.

Since some people must be crushed, we should join with that authority to make sure that it is those we despise, rather than us.

By getting rid of such inferior people, we can preserve scarce resources and advance human social evolution, thereby helping to make the world a better place.

The fact that this case for oppression, tyranny, war, and genocide is entirely false has made it no less devastating. Indeed, it has been responsible for most of the worst human-caused disasters of the past two hundred years. So, let's take it apart.

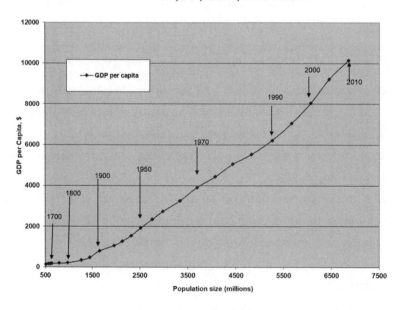

FIGURE 12.3. *Contrary to Malthus's theory, human global well-being has increased with population size, and at an accelerating rate (2020 dollars). (Credit: Author)*

Two hundred years ago, the English economist Thomas Malthus set forth the proposition that population growth must always outrun production as a fundamental law of nature. This theory provided the basis for the cruel British response to famines in Ireland and India during the mid through late nineteenth century, denying food aid or even taxation, rent, or regulatory relief to millions of starving people on the pseudoscientific grounds that their doom was inevitable. It then provided the ideological foundation for much worse atrocities of the twentieth century, notably those of the Nazis.[11]

Yet, the data shows that the Malthusian theory is entirely counterfactual. In fact, over the two centuries since Malthus wrote, world population has risen sevenfold while inflation-adjusted global gross domestic product per capita has increased by a factor of 50, and absolute total GDP by a factor of 350.

Indeed, the Malthusian argument is fundamentally nonsense because resources are a function of technology, and the more people there are, and the higher their living standard, the more inventors, and thus inventions, there will be—and the faster the resource base will expand.

Our resources are growing, not shrinking, because resources are defined by human creativity. In fact, as mentioned previously in this writing, there is no such thing as "natural resources." There are only natural raw materials. It is human ingenuity that turns natural raw materials into resources.

It is for this reason that, contrary to Malthus and all his followers, the global standard of living has continuously gone up as the world's population has increased, not down. The more people,—especially free and educated people—the more inventors, and inventions are cumulative.

Furthermore, the idea that nations are in a struggle for existence is completely wrong. Darwinian natural selection is a useful theory for understanding the evolution of organisms in nature, but it is false as an explanation of human social development. Why? Because, unlike animals or plants, humans can inherit acquired characteristics—for example new technologies—and do so not only from parents but from those to which they are entirely unrelated. Thus, inventions made anywhere ultimately benefit people everywhere. Human progress does not occur by the mechanism of militarily superior nations eliminating inferior nations. Rather, inventions made in one nation are transferred all over the world, where, newly combined with other technologies and different mindsets, they blossom in radical new ways.

Nevertheless, so long as humanity is limited to one planet, the arguments of the Malthusians have the appearance of self-evident truth, and their triumph can only have the most catastrophic results.

Indeed, one has only to look at the history of the twentieth century, and the Malthusian / national social Darwinist rationale that provided the drive to war of both Imperial and especially Nazi Germany to see the horrendous consequences resulting from the widespread acceptance of such myths.

As I write these lines, Russia has invaded Ukraine, which it needs to conquer if it is to consolidate its position to launch a general war.

This is madness. There is no reason for a general war. People all over the world today are living much better than they ever did before, at any time in human history. But the same was true in 1914. As in 1914 and 1939, all it takes is the *belief* that there isn't enough to go around, that others are using too much, or threatening by their growth to do so in the future, to set the world ablaze.

If the New World again rises to the challenge and comes to the rescue of the Old, the current invasion can be repelled. This will stall the Kremlin's plans for aggression for a while.

But let us be clear: We are living on borrowed time. If it is accepted that the future will be one of resource wars,[12] there are men of action who are prepared to act accordingly.

As Adolf Hitler put in in 1941, "The laws of existence require uninterrupted killing, so that the better may live." The putative "fact" of limited resources makes war necessary, which in turn justifies tyranny. Q.E.D.

The problem is much bigger than Putin. People are being bombarded on all sides with propaganda not only by those seeking trade wars, immigration bans, or preparations for resource wars, but by those who, portraying humanity as a horde of vermin endangering the natural order, are promoting an ideology that must ultimately set all against all.

The real lesson of the last century's genocides is this: We are not endangered by a lack of resources. We are endangered by those who believe there is a shortage of resources. We are not threatened by the existence of too many people. We are threatened by people who think there are too many people.

If the twenty-first century is to be one of peace, prosperity, hope, and freedom, a definitive and massively convincing refutation of these pernicious ideas is called for—one that will forever tear down the walls of the mental prison these ideas would create for humanity.

A free Mars would provide a decisive edge to the forces of liberty in the struggles of the future. But it could do even more. It could prevent such wars altogether by undermining the limited resource belief structure that is their driving cause.

Why kill each other fighting over provinces when by working together we can create planets?

A Question of Faith

Science, reason, morality based on individual conscience, and human rights; this is the Western humanist heritage. Whether expressed in Hellenistic, Christian, Deist, or purely Naturalistic forms, it all drives toward the assertion of the fundamental dignity of man. As such, it rejects human sacrifice and is ultimately incompatible with slavery, tyranny, ignorance, superstition, perpetual misery, and all other forms of oppression and degradation. It asserts that humanity is capable and worthy of progress.

This last idea—progress—is the youngest and proudest child of Western humanism. Born in the Renaissance, it has been the central motivating idea of our society for the past four centuries. As a civilizational project to better the world for posterity, its results have been spectacular, advancing the human condition in the material, medical, legal, social, moral, and intellectual realms to an extent that has exceeded the wildest dreams of its early utopian champions.

Yet now it is under attack. It is being said that the whole episode has been nothing but an enormous mistake, that in liberating ourselves we have destroyed the Earth. As influential Malthusians Paul Ehrlich and John Holdren put it in their 1971 book, *Global Ecology*: "When a population of organisms grows in a finite environment, sooner or later it will encounter a resource limit. This phenomenon, described by ecologists as reaching the 'carrying capacity' of the environment, applies to bacteria on a culture dish, to fruit flies in a jar of agar, and to buffalo on a prairie. It must also apply to man on this finite planet."

Note the last sentence: *It must also apply to man on this finite planet.* Case closed. The only thing left to decide is who gets death and who gets jail.

We need to refute this. The issue before the court is the fundamental nature of humankind. Are we destroyers or creators? Are we the enemies of life or the vanguard of life? Do we deserve to be free?

Ideas have consequences. Humanity today faces a choice between two very different sets of ideas, based on two very different visions of the future. On the one side stands the antihuman view, which, with complete disregard for its repeated prior refutations, continues to postulate a world of limited supplies, whose fixed constraints demand ever-tighter controls upon human aspirations. On the other side stand those who believe in the power of unfettered creativity to invent unbounded resources and so, rather than regret human freedom, demand it as our birthright. The contest between these two outlooks will determine our fate.

If the idea is accepted that the world's resources are fixed with only so much to go around, then each new life is unwelcome, each unregulated act or thought is a menace, every person is fundamentally the enemy of every other person, and each race or nation is the enemy of every other race or nation. The ultimate outcome of such a worldview can only be enforced stagnation, tyranny, war, and genocide. Only in a world of unlimited resources can all men and women be brothers and sisters.

On the other hand, if it is understood that unfettered creativity can open unbounded resources, then each new life is a gift, every race or nation is fundamentally the friend of every other race or nation, and the central purpose of government must not be to restrict human freedom, but to defend and enhance it at all costs.

It is for this reason that we need urgently to open the Martian frontier. We must joyfully embrace the challenge of launching new, dynamic, pioneering branches of human civilization on Mars—so that their optimistic, impossibility-defying spirit will continue to break barriers and point the way to the incredible plentitude of possibilities that urge us to write our daring, brilliant future among the vast reaches of the stars. We need to show for all to see in the most sensuous way possible what the great Italian Renaissance humanist Giordano Bruno boldly proclaimed: "There are no ends, limits, or walls that can bar us or ban us from the infinite multitude of things."

That is why we must take on the challenge of Mars. For, in doing so, we make the most forceful statement possible that we are living not at the end of history, but at the beginning of history; that we believe in freedom and not regimentation, in progress and not stasis, in love rather than hate, in peace rather than war, in life rather than death, and in hope rather than despair.

The cause of Mars is the cause of all nations.

EPILOGUE

WHAT NEEDS TO BE DONE NOW?

IF WE WANT TO MAKE NEW WORLDS for humanity on Mars possible in our lifetime, we need to start now. Two of the most essential tasks are to create the technological and economic foundations for Mars city-states. There are two key initiatives that could go a long way toward getting things moving on these fronts. I call these the Mars Technology Institute and the Space Mining Patent Law.

The Mars Technology Institute

SpaceX and other entrepreneurial launch companies are already moving rapidly to develop the transportation systems that can get us to Mars. What is needed is an institution devoted to developing the technologies that will allow us to live once we are there.

Again, there are no such things as natural resources. There are only natural raw materials. It is human creativity, manifested in the form of technology, that transforms materials into resources. We need to start inventing the technologies that will make Mars resource-rich.

All the necessary materials are already there. But we need to develop the necessary science to transform them into fuel and oxygen, food, bricks, cement, metals, glass, fabrics, plastics, energy, and labor power itself. As explained in this book, Mars cities themselves will be inventors colonies devoted to pursuing advances in these areas and will use the resulting technologies both to survive and prosper on Mars, and to obtain the income necessary to pay for imports by licensing such inventions on the home planet.

So, why not start a Mars Technology Institute on Earth now? The business case for investing in such an organization is not as obvious as that for a new launch company, or any other firm devoted to addressing an existing terrestrial market. That's why it hasn't happened already. The initial investors will have to be motivated by a long-term vision, rather than short-term gain. It is Hope, rather than Greed, that will get us to Mars.

But MIT has made a fortune by patenting and licensing the inventions of its scientists. Eventually the MTI will be able to do so, too, and in the process of making a bundle not only create the first generation of technologies that will allow us to get started on Mars but demonstrate the practicality of one of the two main economic foundations for Mars city-states as well.

The Space Mining Patent Law

Inventing and licensing new technologies will be one of the two principal economic pillars of Mars. The other will be the Red Planet's key role as the logistical base supporting the mining of the precious metals that are abundant in the asteroid belt. The surest way to make money in a gold rush is not to mine the gold yourself, but to sell blue jeans to gold miners. Because of its lower gravity and position in the solar system, it will be about two orders of magnitude cheaper to ship necessary supplies and equipment to the asteroids from Mars than from Earth.

What Seattle was to the Yukon, Mars can be to the asteroids.

But for this to happen, there needs to be an asteroidal Yukon.

It is not currently technologically possible to actually mine the asteroids. But we do have the means to explore them. There have been plenty of people on Earth for the past several centuries who made fortunes exploring for gold or oil who never had the ability or intention of mining or drilling for such resources themselves. Rather, the way such prospectors made their cash was simply by searching for resource-rich sites, staking a claim, and then selling that claim to companies prepared to do the digging.

Explorers can make money in this way because there are legal mechanisms for staking such claims. If there were no such mechanisms, no one would search, no one would find, and no one would mine.

We have the technology, today, to prospect asteroids for precious metal resources. If there were a way for prospecting missions to stake legally valid claims, such claims would have speculative value, and extensive, privately funded asteroid exploration could get underway now. Then, once the richest resources were identified, large commercial organizations would not only buy the claims, but develop the heavy-duty technology required to exploit them. In doing so, they would not be expropriating resources that belong to the common heritage of mankind, as some of collectivist persuasions suggest. Rather they would be providing humanity with resources that humanity would not otherwise ever have.

But there is a difficulty. The Outer Space Treaty of 1967 forbids any nation to claim an extraterrestrial object as its sovereign territory. So, the United States, for example, could not claim a given asteroid as its sovereign territory, then grant land or mineral rights there as it does here.

I believe there is a way around this, however, based on patent law. The US Patent and Trademark Office (about which I have mixed feelings since I have had extensive dealings with them) has been of great benefit not just to American inventors, but all of humankind, because it has allowed inventors from every country to turn their creative ideas into negotiable property. Think about it. Let's say you are a Finnish inventor. A Finnish patent is of very little value. Anyone interested in stealing your idea would happily leave the Finnish market to you while they make millions selling your invention to the rest of the world. But whatever country you come from, you can go to the USPTO and get a US patent, giving you exclusive rights to the largest single market in the world. Anyone wishing to produce your invention would be wise to see about cutting a deal with you.

US patents are monopolies granted for eighteen years in return for the inventor fully disclosing and publishing his or her concept. The inventor then has eighteen years in which to reap his or her reward, after which the invention becomes available for implementation by everyone.

So, let's say we had a similar law, and set up a Space Mining Patent Office (SMPO), which granted a mining claim to anyone who surveyed a given asteroid to some specified degree, for example, photographed most of its surface with a resolution of one square centimeter per pixel. The duration of the patent would have to be longer than eighteen years, possibly as long as ninety-nine years, because the technology to exploit such claims will take a while to develop. But you can bet your bottom dollar that they will get developed, once suitably rich claims were found. The Invisible Hand would see to that.

We would not need the Space Force to send patrol vessels to the asteroid belt to enforce SMPO claims any more than we need to send the FBI into foreign countries today to prevent the manufacture of devices in defiance of USPTO patents. Anyone wishing to sell a patented product inside the United States must deal with the authorities here. Whether the platinum from an asteroid is shipped directly into the United States from space, or first goes to a foreign country that puts it into the catalytic converters of cars to be exported here, it must first pass customs. If the material has been mined in defiance of the SMPO patent, it will be treated as contraband. Of course, there are countries like China, where the claim jumper might be able to sell the stuff. But that is true of any patented invention today. Regardless, US patents still have value.

As I see it, there are several ways such an SMPO could be established. It could be done unilaterally by the United States, just as the USPTO was. The only problem with that approach is that Congress is senile, so that getting the required legislation passed could be a heavy lift.

The other approach would be for a small country—for example, Luxembourg, whose government has shown an interest in becoming a significant international player in space—to establish such an office, and then seek international agreements with the United States and other large countries to honor its patents. Having the SMPO set up by a small country that is not perceived as a threat by anyone could facilitate its acceptance by the international community. That is why the Red Cross is based in Switzerland and the International Criminal Court is based in The Hague. It's easier to accept someone as umpire if he or she is not a player.

But one way or another, it is vital that some such institution be established. Things only have value because a property claim can be attached to them. Think of all the houses in the world. None of them, no matter how nice, would be worth a cent if it lacked a legal deed. You don't own your house because you live in it. No matter how individually formidable you may be, a suitable gang of armed men could come in and throw you out—except for the fact that you have a piece of paper that allows you to call upon the police, or if necessary, the army, to protect your claim. All the world's houses, taken together, are worth trillions because there are property rights here. Without property rights, they would be worth zero.

The asteroids can be worth trillions, too, provided the right legal framework is created.

Liberty under law requires laws.

There are still more issues that need to be addressed. The "planetary protection" regulations that NASA has put in place to protect Mars from the entry of Earth life need to be abolished, or human exploration of Mars—let alone settlement—will simply be impossible. Other regulations either already exist or are likely to be proposed that would greatly impede the development, acquisition, or operation of reusable spacecraft, space nuclear reactors, or other critical technologies advanced by private groups need to be prevented or set aside. In short, there needs to be a social consensus that the human settlement of Mars is a desirable goal, or it won't be allowed to happen. The fight to make this positive vision of the human future prevail requires a movement involving the talents of all kinds of people, ranging from politicians, pilots, businessmen, scientists, and engineers, to philosophers, writers, historians, filmmakers, folk singers, artists, and poets.

In 1998 a meeting of seven hundred people was held in Boulder, Colorado, to found such a movement. It's called the Mars Society, and I am its president. In the quarter-century since its founding, it has grown to involve tens of thousands of people from all over the world.

Here is our Founding Declaration:

The time has come for humanity to journey to the planet Mars.

We're ready. Though Mars is distant, we are far better prepared today to send humans to the Red Planet than we were to travel to the Moon at the commencement of the space age. Given the will, we could have our first crews on Mars within a decade.

The reasons for going to Mars are powerful.

We must go for the knowledge of Mars. Our robotic probes have revealed that Mars was once a warm and wet planet, suitable for hosting life's origin. But did it? A search for fossils on the Martian surface or microbes in groundwater below could provide the answer. If found, they would show that the origin of life is not unique to the

Earth, and, by implication, reveal a universe that is filled with life and probably intelligence as well. From the point of view of learning our true place in the universe, this would be the most important scientific enlightenment since Copernicus.

We must go for the knowledge of Earth. As we begin the twenty-first century, we have evidence that we are changing the Earth's atmosphere and environment in significant ways. It has become a critical matter for us better to understand all aspects of our environment. In this project, comparative planetology is a very powerful tool, a fact already shown by the role Venusian atmospheric studies played in our discovery of the potential threat of global warming by greenhouse gases. Mars, the planet most like Earth, will have even more to teach us about our home world. The knowledge we gain could be key to our survival.

We must go for the challenge. Civilizations, like people, thrive on challenge and decay without it. The time is past for human societies to use war as a driving stress for technological progress. As the world moves toward unity, we must join together, not in mutual passivity, but in common enterprise, facing outward to embrace a greater and nobler challenge than that which we previously posed to each other. Pioneering Mars will provide such a challenge. Furthermore, a cooperative international exploration of Mars would serve as an example of how the same joint-action could work on Earth in other ventures.

We must go for the youth. The spirit of youth demands adventure. A humans-to-Mars program would challenge young people everywhere to develop their minds to participate in the pioneering of a new world. If a Mars program were to inspire just a single extra percent of today's youth to scientific educations, the net result would be tens of millions more scientists, engineers, inventors, medical researchers, and doctors. These people will make innovations that *create new industries, find new medical cures, increase income, and benefit the world in innumerable ways to provide a return that will utterly dwarf the expenditures of the Mars program.*

We must go for the opportunity. The settling of the Martian New World is an opportunity for a noble experiment in which humanity has another chance to shed old baggage and begin the world anew; carrying forward as much of the best of our heritage as possible and leaving the worst behind. Such chances do not come often and are not to be disdained lightly.

We must go for our humanity. Human beings are more than merely another kind of animal, we are life's messenger. Alone of the creatures of the Earth, we have the ability to continue the work of creation by bringing life to Mars, and Mars to life. In doing so, we shall make a profound statement as to the precious worth of the human race and every member of it.

We must go for the future. Mars is not just a scientific curiosity; it is a world with a surface area equal to all the continents of Earth combined, possessing all the elements that are needed to support not only life, but technological society. It is a New World, filled with history waiting to be made by a new and youthful branch of human civilization that is waiting to be born. We must go to Mars to make that potential a reality. We must go, not for us, but for a people who are yet to be. We must do it for the Martians.

Believing therefore that the exploration and settlement of Mars is one of the greatest human endeavors possible in our time, we have gathered to found this Mars Society, understanding that even the best ideas for human action are never inevitable, but must be planned, advocated, and achieved by hard work. We call upon all other individuals and organizations of like-minded people to join with us in furthering this great enterprise. No nobler cause has ever been. We shall not rest until it succeeds.

You can join us by visiting our website at www.marssociety.org. I hope you do.

The logo of the Mars Society shows Mars being explored first by robots, with the subsequent arrival of humans eventually resulting in the transformation of the Red Planet into a living world.

Let us make this vision a reality.

ACKNOWLEDGMENTS

I wish to acknowledge the extensive support I have received in my research on Mars in-situ resource utilization technology that I have received over the years by members of the Pioneer Astronautics team, including Mark Berggren, Stacy Carrera, Heather Rose, Steven Fatur, James Siebarth, Tony Muscatello, Douwe Bruinsma, and Brian Frankie. I also wish to acknowledge the extensive debt I owe to members of the forty-two semi-finalist and finalist Mars Colonies and Mars City States design competition teams, who collectively advanced an intellectual banquet of ideas for dealing with virtually every issue bearing on the settlement of the Red Planet, and to Frank Crossman, who organized this enormous wealth of material into two wonderful volumes, from which I could, and did, draw from freely in developing the concepts presented in this book. I also wish to thank my indefatigable literary agent, Laurie Fox, who sold the book, and to my editor at Diversion Books, Keith Wallman, who provided just the right amount of editing. Most of all I wish to thank my wife Hope, without whose loving support this work would not have been possible.

NOTES

CHAPTER 1

1. Robert Zubrin, *The Case for Space* (Amherst, NY: Prometheus Books, 2019).

CHAPTER 2

1. This is a simplified map of Mars based on data produced by the Mars Orbit Laser Altimeter (MOLA) instrument team on the Mars Global Surveyor. A wonderfully detailed map based on MOLA data has been produced by the US Geological Service. It includes the polar regions and many more place names and uses color shading to show the elevation of all locations. It can be found at https://astropedia .astrogeology.usgs.gov/download/Docs/Globes/i2782_sh1.pdf (accessed March 19, 2023).
2. Frank Crossman, *Mars Colonies: Plans for Settling the Red Planet* (Lakewood, CO: Polaris Books, 2019).
3. Frank Crossman, *Mars City States: New Societies for a New World* (Lakewood, CO: Polaris Books, 2020).

CHAPTER 3

1. "Number of Worldwide Launches from 1957 to 2022," Statistica, https://www .statista.com/statistics/1343344/orbital-space-launches-global/ (accessed March 26, 2023).

CHAPTER 4

1. R. L. Ash, W. L. Dowler, and G. Varsi, "Feasibility of Rocket Propellant Production on Mars," *Acta Astronautica*, Vol. 5, July–August 1978, pp. 705–724.
2. B. Frankie and R. Zubrin, "Chemical Engineering in Extraterrestrial Environments," *Chem. Eng. Progress*, 94 (2), pp. 45–54 (February 1999).
3. A. J. Meier, M. Grashik, M. Shah, J. Sass, J. Bayliss, P. Hintze, and R. Carro, "Full-Scale CO_2 Freezer Project Developments for Mars Atmospheric Acquisition." In 2018 AIAA Space and Astronautics Forum and Exposition, 2018 (p. 5172).

4. A. C. Muscatello, P. E. Hintze, A. J. Meier, J. Bayliss, R. Formoso, M. Shah, B. Vu, R. Lee, and J. Captain, "Testing and Modeling of the Mars Atmospheric Processing Module." In AIAA Space and Astronautics Forum and Exposition, 2017 (p. 5149).

5. R. Zubrin, S. Price, L. Mason, and L. Clark, "An End to End Demonstration of Mars In-Situ Propellant Production," AIAA-95-2798, 31st AIAA/ASME Joint Propulsion Conference, San Diego, CA, July 10–12, 1995.

6. Gerald B. Sanders, Aaron Paz, Lara Oryshchyn, Koorosh Araghi, Anthony Muscatello, Diane L. Linne, Julie E. Kleinhenz, and Todd Peters, "Mars ISRU for Production of Mission Critical Consumables—Options, Recent Studies, and Current State of the Art," AIAA 2015-4458, Published Online: August 28, 2015, https://arc.aiaa.org/doi/10.2514/6.2015-4458 (accessed December 29, 2020).

7. Robert Zubrin, Steven Fatur, Heather Rose, and Maxim Shub, "Liquid Sorption Pump," Final Report on NASA SBIR Phase I Contract #: 80NSSC18P1936, Delivered to NASA GRC, January 25, 2019.

8. J. Williams, S. Coons, and A. Bruckner, "Design of a Water Vapor Adsorption Reactor for Martian In situ Resource Utilization," *Journal of the British Interplanetary Society*, August 1995.

9. Mark Berggren, Robert Zubrin, Heather Rose, Anterra Kennedy, and Maxim Shub, "Advanced Mars Water Acquisition System," Phase I Final Report, NASA Contract NASA SBIR Phase I Contract NNX17CC70P, delivered to NASA GRC, December 8, 2017.

10. R. Zubrin and D. Baker, "Mars Direct, Humans to the Red Planet by 1999," IAF-90-672, 41st Congress of the International Astronautical Federation, Dresden, Germany, October 1990. Republished in *Acta Astronautica*, 26, no. 12 (1992): pp.899–912. See also, R. Zubrin, *The Case for Mars: The Plan to Settle the Red Planet and Why We Must*, Simon and Schuster, New York, 1996, 2011.

11. E. Musk, presentation to International Astronautical Congress, Guadalajara, Mexico, September 2016.

12. D. Rapp. "Use of Extraterrestrial Resources for Human Space Missions to Moon or Mars," Chapter 2, Springer Verlag, Chichester, Heidelberg, 2013.

13. Michael H. Hecht and Jeffrey A. Hoffman, "The Mars Oxygen ISRU Experiment (MOXIE) on the Mars 2020 Rover," 3rd International Workshop on Instrumentation for Planetary Missions (2016), https://www.hou.usra.edu/meetings/ipm2016/pdf/4130.pdf (accessed December 29, 2020).

14. R. Zubrin, B. Frankie, and T. Kito, "Mars In-Situ Resource Utilization Based on the Reverse Water Gas Shift," AIAA-97-2767, 33rd AIAA/ASME Joint Propulsion Conference, Seattle, WA July 6–9, 1997.

15. J. Lewis and R. Lewis, "Space Resources: Breaking the Bonds of Earth," Columbia University Press, New York, 1987.

16. R. Zubrin, The Case for Nukes: How We Can Beat Global Warming and Create a Free, Open, and Magnificent Future (Lakewood, CO: Polaris Books), 2023.

CHAPTER 5

1. P. Brisson, R. Heidmann, and T. Volkova, "The Team 'Let It Be' Mars Settlement: How the Utopia Could Materialize," in Frank Crossman, ed, *Mars Colonies: Plans for Settling the Red Planet* (Lakewood: Polaris Books, 2019).

2. Escape velocity on Mars is 5 km/s. With a velocity of 5.8 km/s, a ship could not only escape Mars, but travel back to Earth on a nine-month, minimum energy orbit. The velocity change, or ΔV, of a rocket is given by:

$$\text{Mass ratio} = \exp(\Delta V/V_e) \qquad (1)$$

   Where the mass ratio is the ratio of the wet mass (i.e., dry mass plus propellant) to the dry mass of the rocket and V_e is the exhaust velocity. Plugging in 6 for the mass ratio and 3.62 km/s for V_e, we find that $\Delta V = 6.5$ km/s, which is enough for the return flight, with some margin for gravity losses during takeoff and course correction during flight. If it were desired to return faster, the Starship tanks could be topped off after reaching low Mars orbit using a tanker Starship, in a manner like that planned for sending Starships from Earth to Mars. But this is unnecessary for cargo missions.

3. Bruce Cordell and Steven Gillett, "The Potential Crustal Resources of Mars," Arizona Univ., *Resources of Near-Earth Space:* Abstracts, https://ntrs.nasa.gov/citations/19910016754 (Accessed September 22, 2023).

4. Michel Lamontagne, et al., "Foundation," Chapter 15 in Frank Crossman, ed., *Mars City States: New Societies for a New World* (Lakewood, CO: Polaris Books, 2020).

5. Robert Zubrin with Richard Wagner, *The Case for Mars: The Plan to Settle the Red Planet and Why We Must* (New York: Simon and Schuster, 1996).

6. Since the Starship has a dry mass of 100 tons and a propellant capacity of 1,000 tons, the cargo capacity of a Starship can be calculated from its mass ratio, M, by using the equation: Cargo $+ 100 = 1000/(M-1)$.

7. This is so because high thrust chemical rockets deliver their ΔV virtually instantly— that is in a time that is very short compared to the duration of the transit. This is much more efficient than using low thrust to gradually change an orbit in the course of a trip.

CHAPTER 6

1. Daniel P. Raymer, James French, D. Felix Finger, Arturo Gomez, Jaspreet Singh, Ramlingam G. Pillai, Matheus M. Monjon, Joabe Marcos de Souza, and Aviv Levy, "The Raymer Manned Mars Airplane: A Conceptual Design and Feasibility Study," AIAA 2021–1187, Published Online: January 4, 2021, https://doi.org/10.2514/6.2021-1187 (accessed March 13, 2023).

2. Robert Zubrin, "Methods of Achieving Long Range Mobility on Mars," 28th AIAA/ASME Joint Propulsion Conference, 1992, https://citeseerx.ist.psu.edu/document?repid=rep1&type=pdf&doi=09a96af4a7fce7525e76d5ecd93d148e4fdccc16 (accessed March 13, 2023).

3. R. Zubrin, "Nuclear Rockets Using Indigenous Martian Propellants," AIAA 89-2768, 25th AIAA/ASME Joint Propulsion Conference, Monterey, CA, July 1989, https://arc.aiaa.org/doi/pdf/10.2514/6.1989-2768 (accessed March 13, 2023). See also https://core.ac.uk/download/pdf/42815706.pdf (accessed March 13, 2023).

4. Robert Zubrin, "The Engineering of a Nuclear Thermal Landing and Ascent Vehicle Utilizing Indigenous Martian Propellant," AIP Conference Proceedings 217, 761 (1991); https://doi.org/10.1063/1.40137 (accessed March 13, 2023).

5. Dave Buden, *Nuclear Thermal Propulsion Systems* (Lakewood, CO: Polaris Books, 2011).

CHAPTER 7

1. World Bank, "Arable Land per Person," https://data.worldbank.org/indicator/AG.LND.ARBL.HA.PC (accessed March 15, 2023).

2. Energy Information Agency, "US Energy Consumption by Source and Sector," https://www.eia.gov/energyexplained/us-energy-facts/images/consumption-by-source-and-sector.pdfvb (accessed March 15, 2023) 73.5 quadrillion BTU/year quoted in the note equals 2.35 TW.

3. Johnathan Strickland, "How Much Power Does the Internet Use," How Stuff Works, https://computer.howstuffworks.com/internet/basics/how-much-energy-does-internet-use.htm (accessed March 15, 2023).

4. Xuan Yang, Fredrik Berthold, and Lars A. Berglund, "High-Density Molded Cellulose Fibers and Transparent Biocomposites Based on Oriented Holocellulose," American Chemical Society, https://pubs.acs.org/doi/10.1021/acsami.8b22134 (accessed September 11, 2022).

5. The Linear No Threshold (LNT) methodology posits that any radiation dose is dangerous, bearing a linear fraction of risk of the larger dose, no matter how small it may be. Thus, according to LNT methodology, if 500 rems is judged to represent a fatal dose, 50 rems would represent a 10 percent risk, 5 rems a 1 percent risk, and so forth. This is nonsense because everyone on Earth receives at least a 0.1 Rem dose from background radiation every year, which, according to LNT methodology, would represent a fractional risk of 0.1/500. Multiplied by 8 billion people this would result in a claim of 0.1*8 billion/500 = 1.6 million deaths globally from radiation every year (out of total worldwide deaths from all causes of about 100 million), which bears no relationship to reality. Applied to other toxins, LNT would produce equally nonsensical results. For example, since drinking 50 glasses of wine in one night would almost certainly be fatal, LNT methodology would predict that someone who drinks 1 glass of wine per night for 50 days would die. It would also predict if 50 million Frenchmen drink 1 glass of wine each tonight, 1 million of them would die. LNT methodology has greatly damaged the nuclear industry, as it has served to greatly exaggerate radiation risk. It was designed for that purpose by its inventor, the antinuclear eugenicist H.J. Mueller.

6. Amanda Solaniuk, Leszek Orzechowski et al., "Twardowsky Colony," in Frank Crossman, editor, *Mars Colonies: Plans for Settling the Red Planet* (Lakewood, CO: Polaris Books, 2019).

7. The idea of making use of Roman and medieval vault and arch type construction techniques to build Martian cities was first proposed by Bruce McKenzie. See Bruce McKenzie "Building Mars Habitats Using Local Materials," AAS 87-216 in C. Stoker, ed. *The Case for Mars III*, Volume 74, Science and Technology Series of the American Astronautical Federation, Univelt, San Diego, 1989.

8. Paul Meillon, "Nerio, Underground Capital of Mars," in Frank Crossman, editor, *Mars City States: New Societies for a New World* (Lakewood, CO: Polaris Books, 2019).

9. Stefanie Schur, "Fossae Colony One: A Design for a Vibrant Mars Colony," in Frank Crossman, editor, *Mars Colonies: Plans for Settling the Red Planet* (Lakewood, CO: Polaris Books, 2019).

10. "Lava Tube," Wikipedia, https://en.wikipedia.org/wiki/Lava_tube (accessed September 11, 2022).

11. Ian O'Neill, "Seventh Graders Discover Martian Cave," Discovery News, June 18, 2010, https://web.archive.org/web/20140929185101/http://news.discovery.com/space/seventh-graders-discover-martian-cave.htm (accessed September 11, 2022).

12. Robert Zubrin, "Sublake Settlements on Mars," *Centauri Dreams*, May 29, 2020, https://www.centauri-dreams.org/2020/05/29/sublake-settlements-for-mars/ Accessed September 11, 2022.

13. Frank Crossman, ed, *Mars Colonies: Plans for Settling the Red Planet* (Lakewood, CO: Polaris Books, 2019).

14. Frank Crossman, ed. *Mars City States: New Societies for a New World* (Lakewood, CO: Polaris Books, 2020).

15. Jane Jacobs, *The Death and Life of Great American Cities* (New York: Random House, 1961).

16. Jeffrey Greenblatt and Akhil Rao, "K-Town: A Thousand Person Colony," in Frank Crossman, editor, *Mars Colonies: Plans for Settling the Red Planet* (Lakewood, CO: Polaris Books, 2019).

17. Matt Wise, Kyle Saffel, Patrick Fagin Douglas Livermore, and Kaitlin Davis, "The Team Bold Mars Colony," in Frank Crossman, editor, *Mars Colonies: Plans for Settling the Red Planet* (Lakewood, CO: Polaris Books, 2019).

18. R. Mahoney, et al., "The Engineering, Economics, and Ethics of a Martian City State," in Frank Crossman, editor, *Mars City States: New Societies for a New World* (Lakewood, CO: Polaris Books, 2019).

19. George Lordos and Alexandros Lordos, "Star City, Mars," in Frank Crossman, editor, *Mars Colonies: Plans for Settling the Red Planet* (Lakewood, CO: Polaris Books, 2019).

20. Muhammad Akbar Hussain, "An Introduction to Engineering, Economic, and Social Perspective of Establishing Human Civilization on Mars," in Frank Crossman,

editor, *Mars City States: New Societies for a New World* (Lakewood, CO: Polaris Books, 2019).

21. Guillem Anglada-Escude et al, "The Sustainable Offworld Network: Nuwa," in Frank Crossman, editor, *Mars City States: New Societies for a New World* (Lakewood, CO: Polaris Books, 2019).

22. Michel Lamontagne et al, "Foundation," in Frank Crossman, editor, *Mars City States: New Societies for a New World* (Lakewood, CO: Polaris Books, 2019).

23. Alex Sharp, Saleem Ameen, and John Walker, "Korolev Crater Special Administrative Region," in Frank Crossman, editor, *Mars City States: New Societies for a New World* (Lakewood, CO: Polaris Books, 2019).

24. Adrian Moisa et al, "Nexus Aurora—Mars City State Design," in Frank Crossman, editor, *Mars City States: New Societies for a New World* (Lakewood, CO: Polaris Books, 2019).

CHAPTER 8

1. John Heinrichs, *The WEIRDest People in the World: How the West Became Psychologically Peculiar and Particularly Prosperous* (New York: Farrar Straus and Giroux, 2020), 97.

2. Alexis de Tocqueville, *Democracy in America*, transl. George Lawrence (New York: Perennial Classics, 2000), 603.

3. Ibn Warraq, *Why I Am Not a Muslim* (Amherst, New York: Prometheus Books, 1995).

4. It is estimated that because of the one-child policy, more than one hundred million Chinese girls were either aborted or murdered in the "dying rooms" of Chinese "orphanages." See Robert Zubrin, *Merchants of Despair* (New York: Encounter Books, 2012).

5. The soaring rates of violent crime in China following the coming of age of excess males are reported in John Heinrichs, *The WEIRDest People in the World.*

6. See John Heinrichs, *The WEIRDest People in the World.* Of all human societies known to anthropology, approximately 85 percent are polygamous, 15 percent are monogamous, and fewer than 1 percent are polyandrous.

7. See Toby Huff, *The Rise of Early Modern Science: Islam, China, and the West* (Cambridge, UK: Cambridge University Press, 1993).

8. See, for example, the Koran, 2:191, "Kill the unbelievers wherever you find them." Jewish scripture argues otherwise; see Leviticus 19 33–34. Informally, however, it is understood within the tribe that "someone has to pay retail." Even Christianity, founded on the instruction that all people love one another, is frequently understood by its followers as limiting such sympathy only toward other Christians, or even others of the same denomination.

9. My guess is that this stems from the fact that since the advent of rock music in the 1960s, popular songs have become increasingly unsingable, and therefore relegated to performance art. Broadway music has also, for the most part, become much less singable since the 1970s. Compare the songs in Steven Sondheim's 1960s musical

West Side Story with those of his 1990s work, *Into the Woods.* Or compare those in *Oklahoma!, South Pacific, My Fair Lady, Guys and Dolls, Fiddler on the Roof,* or *Man of La Mancha* with *Cats.* Case closed. Classical music still exists, of course, but the snobs who control the classical music scene have kept it out of popular culture by refusing to have operas performed in English.

CHAPTER 9

1. Robert Zubrin, *The Case for Nukes: How We Can Beat Global Warming and Create a Free, Open, and Magnificent Future* (Lakewood, CO: Polaris Books, 2023).

2. Robert G. Ingersoll, "Indianapolis Speech," 1876, see https://infidels.org/library /historical/robert-ingersoll-indianapolis-speech76/ (accessed November 27, 2021). The suggestion that space colonies might tend toward tyranny is argued in Charles Cockell, *Extra-Terrestrial Liberty: An Enquiry Into the Nature and Causes of Tyrannical Government Beyond the Earth* (London: Shoving Leopard, 2013). In his subsequent book, *Interplanetary Liberty: Building Free Societies in the Cosmos* (Oxford, UK: Oxford University Press, 2022), Cockell refined this view to argue that all governments have in their nature tyrannical behavior. All governments are potentially benevolent as well, but an overabundance of benevolence is rarely something we have to concern ourselves with when it comes to the state. Therefore we should assume that people will behave badly and then build government accordingly, with suitable checks and balances. The interdependence in space (such as oxygen, water, etc.) gives extra levers to the tyrant. Therefore, says Cockell, we should start by assuming the worst (that they will be used) and plan accordingly to ensure the best outcome.

3. Thucydides, *History of the Peloponnesian War*, transl. Rex Warner (New York: Penguin, 1972).

4. See, for example Kent Nebergall, "The Eureka Design," in Frank Crossman, ed, *Mars Colonies: Plans for Settling the Red Planet* (Lakewood, CO: Polaris Books, 2019).

5. Crossman, ed. *Mars City States: New Societies for a New World* (Lakewood, CO: Polaris Books, 2020).

CHAPTER 10

1. James McCabe Jr. *Great Fortunes and How They Were Made* (Philadelphia: George MacLean, 1871). Available from Project Gutenberg, https://www.gutenberg.org /files/15161/15161-h/15161-h.htm (accessed March 21, 2023).

2. Building his first western steamboat in Pittsburgh, Nicholas Roosevelt (great uncle of President Theodore Roosevelt) took command and, taking his wife, the former Lydia Lathrop, with him, set forth down the Ohio River on the *New Orleans* bound for the city of the same name in October 1811. The voyage was an epic adventure. With the great comet of 1811 shining overhead, the *New Orleans* sailed through the

devastating New Madrid earthquake of 1811 (the largest in Midwestern history, registering 8.0 on the Richter scale, and ringing church bells as far as Boston), amid the raging frontier war with Tecumseh's British-backed tribal confederation. The earthquake lowered the river's water level and filled its channels with uprooted trees and other debris, forcing the *New Orleans* to make an extended stop in Louisville, during which Lydia had a baby and Tecumseh was beaten at Tippecanoe. Then, on November 30, the river was judged high enough to proceed and steamed off at full speed, at one point escaping Chippewa Indians in hot pursuit in war canoes. The *New Orleans* reached Natchez by December 30 and New Orleans just ten days later. The *New Orleans* did not have the power to steam up the Mississippi beyond Natchez, so Roosevelt left her in the South, where it served traffic on the lower Mississippi for the next several years. It was quickly joined by many others, and three years later took part in the steamer flotilla that delivered critical arms and ammunition to Andrew Jackson's army right before the Battle of New Orleans.

3. Robert Zubrin with Richard Wagner, *The Case for Mars, the Plan to Settle the Red Planet and Why We Must,* Chapter 8 (New York: Simon and Schuster, 1996, 2011, 2021).

4. Robert Zubrin, *The Case for Space: How the Revolution in Spaceflight Open Up a Future of Limitless Possibility,* Chapter 5 (Amherst: Prometheus Books, 2021).

5. Progress in fusion energy is measured by the Lawson triple produce, which is the plasma density times temperature, times confinement time.

6. Robert Zubrin, *The Case for Nukes: How We Can Beat Global Warming and Create a Free, Open, and Magnificent Future* (Lakewood, CO: Polaris Books, 2023).

7. Instead of using exterior magnetic fields to confine a plasma, certain fusion devices drive currents within the plasma to create closed magnetic field lines within it. The most prominent of such systems are the Field Reversed Configurations, or FRCs, which work by creating plasma "smoke rings" inside a cylindrical vessel. These systems routinely achieve plasma pressures that are about half that of the magnetic pressure created by the exterior magnets surrounding the cylindrical vacuum vessel. In contrast, the toroidal-shaped tokamaks that dominate current government fusion budgets only achieve plasma pressures that are about 10 percent that of the confinement field. I believe that FRCs are the most promising concept for both fusion rockets and commercial fusion reactors. They need to be researched much more vigorously.

8. A rocket can generally be engineered to achieve a ΔV up to about twice its exhaust velocity. So, a fusion rocket with an exhaust velocity of 6 percent light speed could potentially reach about 12 percent the speed of light. This would allow it to reach the closest star, 4.2 light-years distance, in forty years, if it did not need to use its rocket engine to stop. This might be possible if a non-rocket deceleration system, like a magnetic sail creating drag against the interstellar medium, were available. Otherwise, the trip could be done in eighty years, with half the ΔV being used to accelerate and half to decelerate.

9. Higher exhaust velocity decreases required propellant mass, and since, for any rocket, the ratio of its wet mass (with propellant) to its dry mass (without propellant) goes in proportion to $e^{\Delta V/v}$, it is very important to have an exhaust velocity, v, that is at least comparable to the mission ΔV. But for any rocket engine of a given power, thrust is inversely proportional to exhaust velocity.

10. "Fusion Torch," Wikipedia. https://en.wikipedia.org/wiki/Fusion_torch (accessed November 26, 2022).

11. Proton-Boron (p-B[11]) fusion requires plasma temperatures over 100 keV, which is impossible to obtain using any currently known exterior plasma heating methods. However, deuterium fusion can be ignited at a readily attainable 5 keV. If the device's confinement time is more than that needed to sustain ignition, its temperature will rise to over 50 keV, at which point deuterium-deuterium fusion will catch fire. This can carry the temperature over 100 keV, enabling p-B[11] fusion to take off.

12. Shannon Nangle, et al., "The Case for Biotech on Mars," *Nature Biotechnology*, 38, 401–407, April 1, 2020, https://www.nature.com/articles/s41587-020-0485-4 (accessed November 26, 2022).

CHAPTER 11

1. In a much-publicized paper published in 2018, Professor Bruce Jakosky, the principal investigator of NASA's MAVEN orbiter, claimed that its results showed that at most Mars could have about 100 mb of CO_2 available for outgassing, and that therefore the Red Planet could not be terraformed. These claims had no basis. MAVEN measured the rate at which Mars is currently losing CO_2 to space. It took no measurements of Mars's regolith or polar CO_2 inventory. In fact, if anything, MAVEN's results suggested that Mars currently has a very large CO_2 inventory. This is because, according to Jakosky's analysis, MAVEN data indicates that Mars has lost about 500 mb of CO_2 to space over the past four billion years. In fact, however, for Mars to have been warm enough for liquid water under the weaker sunlight of three to four billion years ago, it would have needed to have a CO_2 atmosphere thicker than 2,000 mb. This would leave an inventory of 1,500 mb still on the planet today.

2. Under Martian temperature conditions, zeolites routinely absorb up to 10 percent of their weight in CO_2. A reasonable guess is that Martian regolith might do half as well.

3. Robert Zubrin, *Entering Space: Creating a Spacefaring Civilization* (New York: Tarcher Putnam, 1999), 170–172.

4. Robert Zubrin with Richard Wagner, *The Case for Mars: The Plan to Settle the Red Planet and Why We Must*, Chapter 9 (New York: Simon and Schuster), 1996, 2011.

5. Robert Forward, "The Statite: A Non-Orbiting Spacecraft," AIAA 89-2546, AIAA/ASME 25th Joint Propulsion Conference, Monterey, CA, July 1989.

CHAPTER 12

1. Frederick Jackson Turner, *The Frontier in American History* (New York: Henry Holt, 1920).

2. Winston Churchill, "We shall fight on the beaches," International Churchill Society, https://winstonchurchill.org/resources/speeches/1940-the-finest-hour/we-shall -fight-on-the-beaches/ (accessed December 3, 2022).

3. I am not a believer in supernatural beings, and consequently do not subscribe to any religion. However, the influence of religions and the philosophical concepts that they carry are very real. In particular, the Judeo-Christian ideas of the human mind as image of God, and humans as children of God, have been of decisive historical importance for the development of Western civilization.

4. American action to rescue Britain did not wait till Pearl Harbor. Within a week of Dunkirk, FDR organized the shipment of one million rifles, seven thousand field artillery pieces, and one hundred million rounds of ammunition to Britain. See Winston Churchill, *Their Finest Hour* (New York: Houghton Mifflin, 1949). In fact, America stopped Europe's decent into darkness not once, but three times—in World War I, World War II, and the Cold War, and as I write these lines, America is, by standing with Ukraine, playing the central role in defending Europe from barbarism today.

5. Don H. Doyle, *The Cause of All Nations: An International History of the Civil War* (New York: Basic Books, 2014). The support of European liberals for the Union was critical. If either Britain or France had entered the war on the side of the slaveholders, the Union would have lost. Mobilization of public opinion by European liberals prevented that disaster. Not only that, but many also chose to fight. Some 40 percent of the North's army were immigrants or first-generation Americans. These included Lincoln's core supporters and the Union's best shock troops, the German American Wide Awakes of the Midwestern frontier. In contrast, only 3 percent of the Confederate forces were drawn from immigrants.

6. The organizers of anti-Columbus agitation say that they are motivated by their need to show opposition to the violence inflicted on Caribbean natives by Columbus and his sailors following their arrival. But as such—and much larger—episodes of inter- and intra-cultural violence were routine in both the Old World and the New at that time, and as the so-called "Wokeists" exhibit little interest in acting to stop far worse atrocities going on in their own day—for example in Syria, the Congo, Yemen, Afghanistan, Ukraine, China, etc.—it is rather clear that the rough edge of the fifteenth-century sea captain's character is not the driver of the current mobilization against him. On the contrary, it is not the rather ordinary negative deeds of Columbus that mandate his "cancelation," but the extraordinary positive one: that he was the person who, through his courage, tenacity, and vision, discovered a New World for the West, thereby making America possible. Columbus spent two decades trying to find sponsorship for his voyage of exploration to the West. He reportedly traveled to Iceland where he could have heard sagas of the Viking discovery of

Vineland. Columbus was also a believer in Atlantis. I think that's what he was really looking for, and the promise of a possible trade route to India was just bait the Italian navigator used to obtain funding from the unsophisticated bumpkins of the Spanish Court. As a loyal Genoese, the last thing he would have wanted would be to provide the Spanish with a trade route that would circumvent that of his homeland. Furthermore, the key concession that Columbus demanded from the Spanish Crown as his payment was to be made governor of any new lands he discovered. In his copy of Seneca's play *Medea*, Columbus underlined the passage where the sorceress Medea says, "The day will come when the sea will give up its secrets and the hidden lands to the west will be revealed to the minds of men." The book was passed on to Columbus's son Ferdinand, who proudly commented in the margin, "This prophesy was realized by my father, the admiral Christopher Columbus, in the Year of our Lord 1492."

7.   Athena and Liberty are related goddesses. As the ancient Greek Stoic philosopher Chrysippus put it, "Only the wise can be free." The statue of Athena on top of the US Capitol Building is known as "The Statue of Freedom," https://en.wikipedia.org/wiki/Statue_of_Freedom (accessed April 15, 2023).

8.   Robert Zubrin, *Merchants of Despair* (New York: New Atlantis Books, 2012).

9.   Rose Wilder Lane, *The Discovery of Freedom: Mans Struggle Against Authority* (Baltimore: Laissez Faire Books, reprint edition, 2012).

10. It is true that most of the German army was destroyed by the Soviets. The Red Army was only able to do so, however, with the help of American equipment, which included not only weapons, but essential dual-use gear. For example, 90 percent of the locomotives that provided the essential transport for the Red Army and its supplies were made in America. Furthermore, it was the British Empire and American forces who cut off German access to the world market and destroyed the Nazi Luftwaffe and Germany's industrial base, including virtually all its capacity for oil production, thereby tilting the odds of eastern front battlefield victory decisively in the Soviets' favor. And if the Soviets had collapsed, we still would have won, using atomic bombs invented by a combination of British and American talent and that of refugee scientists fleeing Hitler's horrid Third Reich.

11. Robert Zubrin, *Merchants of Despair* (New York: New Atlantis Books, 2012).

12. Michael Klare, *Resources Wars: The New Landscape of Global Conflict* (New York: Macmillan, 2001).

INDEX

ABOUT THE AUTHOR

DR. ROBERT ZUBRIN is President of The Mars Society and a veteran astronautical engineer. In 1996, he founded Pioneer Astronautics, an aerospace R&D company he led for twenty-seven years, successfully executing more than seventy programs for NASA, the US Air Force, and the Department of Energy until selling the firm in 2023. Prior to founding Pioneer Astronautics, Dr. Zubrin worked as a Senior Engineer at Martin Marietta and Lockheed Martin, as well as in the areas of nuclear power plant safety, radiation protection, and thermonuclear fusion research. He holds a Master's degree in Aeronautics and Astronautics and a PhD in Nuclear Engineering from the University of Washington. He is a Fellow of the British Interplanetary Society, the former Chairman of the Executive Committee of the National Space Society, and the inventor of twenty patents. Dr. Zubrin is the author of more than two hundred published technical and nontechnical papers in the field of space exploration and technology, and twelve books, including *The Case for Mars: The Plan to Settle the Red Planet and Why We Must*, now in its 25th Anniversary Edition. He lives with his wife, Hope, a retired science teacher, in Golden, Colorado.